XI Bulgarian-Serbian Astronomical Conference (XI BSAC)

Publications of the Astronomical Society "Rudjer Boskovic"
Number 18

Belogradchik, Bulgaria
14-18 May 2018

Editors:

Milcho K. Tsvetkov
Momchil Dechev
Milan S. Dimitrijevic

ISBN: 978-1-5108-7972-0

Printed from e-media with permission by:

Curran Associates, Inc.
57 Morehouse Lane
Red Hook, NY 12571

CURRAN ASSOCIATES INC.
proceedings
.com

Some format issues inherent in the e-media version may also appear in this print version.

Printed by Curran Associates, Inc. (2019)

For permission requests, please contact Serbian Astronomical Society
at the address below.

Serbian Astronomical Society
Volgina 7
11000 Belgrade
Serbia

Phone: +381-11-3089-072

mdimitrijevic@aob.bg.ac.rs

Additional copies of this publication are available from:

Curran Associates, Inc.
57 Morehouse Lane
Red Hook, NY 12571 USA
Phone: 845-758-0400
Fax: 845-758-2633
Email: curran@proceedings.com
Web: www.proceedings.com

ПУБЛИКАЦИЈЕ АСТРОНОМСКОГ ДРУШТВА "РУЂЕР БОШКОВИЋ"
PUBLICATIONS OF THE ASTRONOMICAL SOCIETY "RUDJER BOŠKOVIĆ"
Св. 18 No. 18

PROCEEDINGS OF THE XI BULGARIAN-SERBIAN ASTRONOMICAL CONFERENCE

Belogradchik, Bulgaria, May 14-18, 2018

Eds. Milcho K. Tsvetkov, Milan S. Dimitrijević and Momchil Dechev

БЕОГРАД
2018

PUBL. ASTRON. SOC. "RUDJER BOŠKOVIĆ" No. 18, 1-254 BELGRADE, NOVEMBER 2018

SCIENTIFIC COMMITTEE

Dr Milcho K. Tsvetkov (Co-chairman)
Dr Milan S. Dimitrijević (Co-chairman)

Dr Momchil Dechev (Vice co-chairman)
Dr Zoran Simić (Vice co-chairman)

Prof. Dr Evgeni Semkov
Prof. Dr Renada Konstantinova-Antova
Prof. Dr Rumen Bachev
Prof. Dr Tanju Bonev
Prof. Assoc. Dr Ivanka Stateva
Prof. Assoc. Dr Antoaneta Antonova
Dr Peter Duchlev
Dr Nikola Petrov
Prof. Dr Dragomir Marchev
Dr Goran Damljanović
Dr Edi Bon
Dr Nataša Bon
Dr Zorica Cvetković
Dr Dragana Ilić
Dr Predrag Jovanović
Prof. Dr Andjelka Kovačević
Dr Jelena Kovačević Dojčinović
Dr Vladimir Srećković
Prof. Dr Dejan Urošević
Dr Sonja Vidojević
Dr Nikolaj Samus
Dr Jan Vondrák
Prof. Dr Branko Predojević

LOCAL ORGANIZING COMMITTEE

Dr Momchil Dechev (Chairman)

Scientific secretary: Dr Kostadinka Koleva

Members:
Svetlana Boeva
Mirela Napetova
Ljuba Vasileva
Georgi Latev

Under the auspices of
Bulgarian Academy of Sciences
Ministry of Education, Science and Technological Development of Serbia

ORGANIZERS:
Institute of Astronomy and National Astronomical Observatory, BAS, Bulgaria
Astronomical Observatory Belgrade, Serbia

Text arrangement by computer: Tatjana Milovanov

President of the Astronomical Society "Rudjer Bošković": Miodrag Dačić

Financially supported by the Ministry of Education, Science and Technological Development of Serbia

ISBN 978-86-89035-11-7

Production: Skripta Internacional, Mike Alasa 54, Beograd, in 100 copies

CONTENTS

Poster Papers

Invited Lectures

Proceedings of the XI Bulgarian-Serbian Astronomical Conference
(XI BSAC) Belogradchik, Bulgaria, May 14-18, 2018
Editors: Milcho K. Tsvetkov, Milan S. Dimitrijević and Momchil Dechev
Publ. Astron. Soc. "Rudjer Bošković" No 18, 2018, 9-12

BLAZAR OPTICAL VARIABILITY: 20 YEARS OF OBSERVATIONS AT BELOGRADCHIK OBSERVATORY

RUMEN BACHEV

Institute of Astronomy and NAO, Bulgarian Academy of Sciences
E-mail: bachevr@astro.bas.bg

Abstract. Blazars are a type of active galactic nuclei with very unusual and interesting properties, one of which is their rapid variability. We show some examples of blazar variability, present a plausible explanation and review the contribution of Belogradchik observatory in blazar studies.

1. INTRODUCTION

The term "blazar" has been introduced about 40 years ago and indicates a special type of active galactic nucleus (AGN), whose emission is first dominated by non-thermal processes in a relativistic jet and secondly – this jet happens to be pointed within a small angle to the line of sight. Thus, due to the relativistic effects, the motion of the jet components often appears "superluminal", as observed from the Earth.

Unlike the other types of AGN, the blazar spectral energy distribution (SED) often covers up to 17-18 orders in magnitude in frequency – from radio to VHE gamma rays. The SED consists typically of two peaks – a synchrotron one (IR/optical regions), due probably to synchrotron emission from relativistic particles and an Inverse Compton (IC) one (X/gamma range), for which the most plausible explanation is inverse Compton scattering of the synchrotron or external photons off the same relativistic electrons. Another option for the high energy peak is the proton-synchrotron instead of the IC process, which however requires a much stronger magnetic field in the jet (10-100G vs. about 0.1G). In addition, some opticl/UV contribution from an accretion disk is also detected in some objects (flat-spectrum radio quasars, FSRQ). Except for the emitted electromagnetic radiation, the blazars were very recently suspected to be also neutrino sources, when a 290TeV neutrino signal was detected from a position, coincident with a blazar (TXS 0506+056), IceCube collaboration (2018a, b).

Blazars are long known to be highly and rapidly variable throughout their entire SED. Such variability is especially well documented in the optical region. The most extreme cases show variations of up to 6 magnitudes on long-term scales and up to 0.3 – 0.7 magnitudes for several hours. Some object show variations of up to 0.2 magnitudes for an hour, which is extreme variability even for a stellar-mass object.

This paper reviews the Belogradchik observatory contribution to the blazar optical variability studies.

2. EXAMPLES OF EXTREME BLAZAR VARIABILITY AND ITS POSSIBLE EXPLANATIONS

In the next figure (Fig. 1) we show examples of long-term optical variability of 2 objects – CTA 102, a FSRQ and BL Lac – the architype of the BL Lac type objects. Most of the data is obtained with Belogradchik 60cm telescope. It has

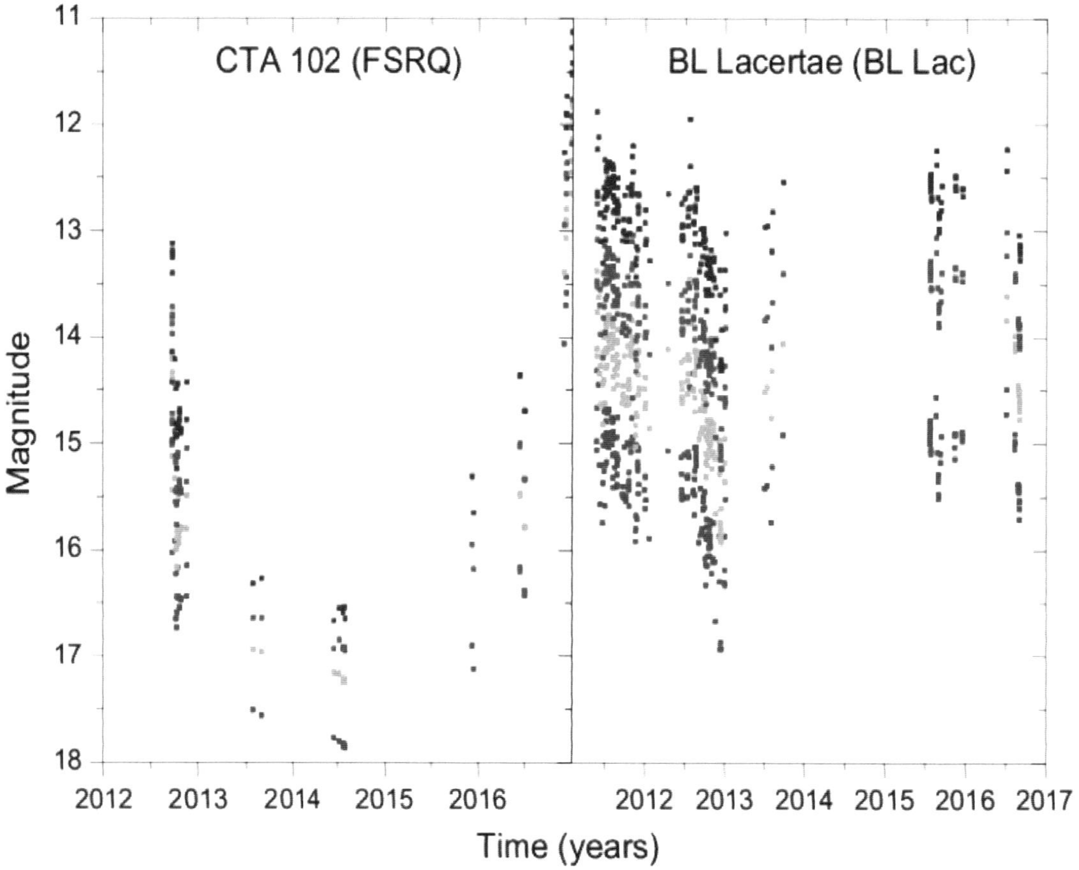

Figure 1: Long-term optical variability of CTA 102 (left) and BL Laceratae (rights). BVRI light curves are shown in blue, green, red and brown, respectively. Most of the data is from Belogradchik 60cm telescope.

already been noticed (Bachev, 2018) that the long-term variations may be characteristically different for the two blazar subclasses: The FSRQ show strong but relatively rare outbursts, while BL Lacs show intermittent but limited in amplitude variations.

On the other hand FSRQs tend to be more variable on intra-night time scales, sometimes reaching changes of about 0.5 – 0.7 mag for several hours. The next example (Fig. 2) shows such rapid variability during an outburst of the blazar CTA 102.

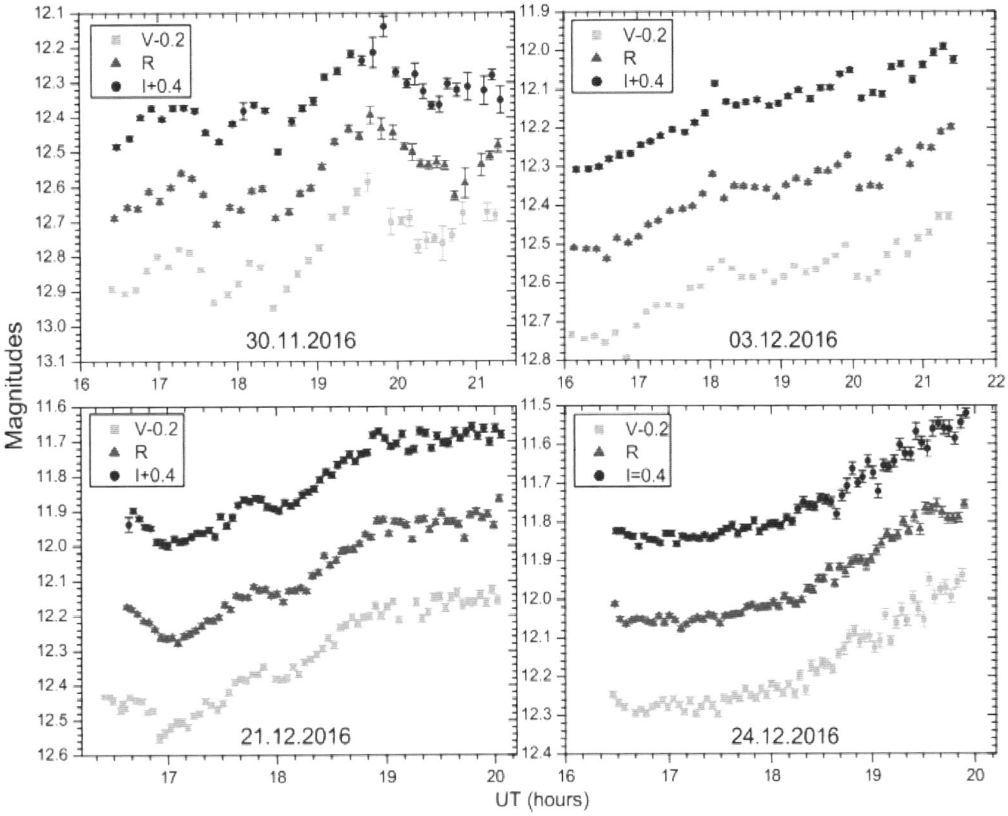

Figure 2: Rapid intra-night variability in 3 colors of CTA 102 during the 2016/2017 outburst (Bachev et al., 2017). Data are obtained from Belogradchik observatory.

The question is how to explain such large-amplitude (on long-term scales) and rapid (on intra-night scales) variations in blazars. There are two possibilities – one related to the evolution of the electron energy density distribution and the other one – related to the change of the Doppler factor in a curved (helical) jet. The first option predicts however some time asymmetry of the light curves as well as some time delays between light curves of different colors. None of these has been positively confirmed throughout the years (Bachev et al., 2017). Therefore, it is likely that a changing Doppler factor (D) plays the most important role in driving blazar variability (Raiteri et al., 2017).

Indeed, knowing that:

$$D = \frac{1}{\Gamma(1-\beta cos\theta)} \quad \text{and} \quad F_{obs} \propto D^{3+\alpha}$$

where $\alpha \approx 1, \Gamma \sim 15$ - 20, is the Lorentz factor, and θ is the jet direction with respect to the line of sight, one sees that a small change in θ can lead to a huge change in the observed flux (F_{obs}). Such light curves will be time-symmetric and no time lags between optical bands should be expected, as observed.

3. BELOGRADCHIK OBSERVATORY AND BLAZAR STUDIES

Belogradchik observatory is situated in the vicinity of the town of Belogradchik, north-west Bulgaria, at 620m above the sea level. The main instrument is a 60cm Cassegrain telescope. The cameras in use are SBIG ST-8 (1997-2009) and FLI PL-9000 (2010-now), both equipped with standard BVRI filters. About 150 clear nights per year are available, of which about 75 are during the non-moon period.

Blazar observations have been performed with this telescope since 1998. Since about 2009 blazar variability studies became the main objective of the Observatory as for them the most of the observing time was dedicated. Throughout the years we have performed over 1000 single BVRI estimates for more than 50 blazars. In addition several object ware monitored on intra-night time scales for a total of more than 500 hours. The most of the observations have been parts of wider international programs for blazar monitoring, such as WEBT, GASP, MAGIC or have been collaborations with groups from India, Poland, Russia, etc. There are more than 40 high-ranking journal (ApJ, A&A, MNRAS, Nature) publications since 2007 on blazar variability, whose data are partially or entirely obtained from Belogradchik.

Acknowledgements

This work was partially supported by the Bulgarian National Science Fund of the Ministry of Education and Science under grants DN 08-1/2016 and DN 18-13/2017.

References

Bachev, R., Popov, V., Strigachev, A., et al.: 2017, *MNRAS* **471**, 2216.
Bachev, R.: 2018, *Bulg. AJ,* **28**, 22.
IceCube collaboration: 2018a, *Science,* **361**, 147.
IceCube collaboration: 2018b, *Science,* **361**, 1378.
Raiteri, C. M., Villata, M., Acosta-Pulido, J. A., Agudo, I., Arkharov, A. A., Bachev, R., et al., 2017, *Nature,* **552**, 374.

Proceedings of the XI Bulgarian-Serbian Astronomical Conference (XI BSAC) Belogradchik, Bulgaria, May 14-18, 2018
Editors: Milcho K. Tsvetkov, Milan S. Dimitrijević and Momchil Dechev
Publ. Astron. Soc. "Rudjer Bošković" No 18, 2018, 13-22

FIRST GAIA DATA RELEASE – DR1 AND SERBIAN-BULGARIAN ASTRONOMICAL ACTIVITIES

GORAN DAMLJANOVIĆ

Astronomical Observatory, Volgina 7, 11060 Belgrade, Serbia
E-mail: gdamljanovic@aob.rs

Abstract. The ESA Gaia mission is surveying the full sky since July 2014 (useful data, from August 2014). It is astronomically, photometrically and spectroscopically European project. Also, it is revolution in astronomy, our understanding of the Milky Way galaxy, stellar physics, the Solar system bodies, etc. At September 2016, the first Gaia data release – DR1 has been made available using about 14 months of Gaia data. At April 2018, the second Gaia solution – DR2 appeared based on 21 months of useful Gaia observations. At the other hand, it is of importance the synergy between Gaia and ground-based astronomical observations. Because of it, the regional astronomical cooperation "Serbian-Bulgarian mini-network telescopes" was started at 2013. The main results of the Gaia DR1 (also, DR2) and mentioned cooperation are presented, here.

1. INTRODUCTION (EXPECTATIONS OF GAIA MISSION)

The GAIA (Global Astrometric Interferometer for Astrophysiscs) is a cornerstone mission of European Space Agency - ESA. The spacecraft (see Fig. 1) was launched at the end of 2013, and the start of normal operations was in July 2014, but useful data for DR2 – from August 2014. It is doing the astronomical observations during five-year nominal operations phase. Now, there is information that it will be longer than predicted five-year period.

The plan was to observe of about one billion objects in our Milky Way galaxy (astrometry, photometry and spectroscopy) and about 500 000 extragalactic sources (Prusti 2012). The goal was to produce the Gaia Catalogue in optical domain and to replace the International Celestial Reference Frame - ICRF (which is based on VLBI radio observations). Also, to measure positions with extremely precision, parallaxes at the level of accuracy far way from the ground ones, photometry for all astrometrically detected objects, in spectroscopy to get the catalogue of radial velocities for near 150 million sources, etc.

There are a lot of open questions related to our Milky Way galaxy (the structure and dynamics of the Galaxy, star formation history, etc.) which could be solved using Gaia data because Gaia is scanning the full sky with very high precision. Gaia data open many more science areas to study and explore.

In stellar astrophysics, the improvement in distances (using the improvement of parallaxes via Gaia observations) will allow to models of stars at different steps of evolution.

In the domain of binary and multiple stars, the orbital parameters could be calculated with much higher precision using the Gaia high spatial resolution.

About the exoplanets, it is expected to find several thousand systems.

In case of faint rare objects (as brown or white dwarfs), Gaia will be of importance.

Mostly, Gaia is done for stars (point objects), but also it is possible to detect solar system objects, galaxies and quasars (QSOs). It is going a massive improvement of asteroid ephemeredes, to determine the masses of the objects in some cases, etc.

It is expected to detect the few million galaxies. The QSOs are of importance for reference systems and fundamental physics.

The HIPPARCOS (High Precision PARallax Collecting Satellite), the predecessor of Gaia, with about 118 000 stars changed the astronomy at the end of last century (ESA 1997, van Leeuwen 2007), but Gaia with about 1 billion stars (and near 500 000 extragalactic sources), from milliarcsec to microarcsec astrometry (plus the radial velocities), will be a revolution in astronomy (Prusti 2012).

Two Gaia fields of view are separated by a basic angle of 106.5 degrees, and Gaia rotates around itself with a period of 6 hours; the period of the precession spin axis is 63 days. For each observed object, the plan is to collect between 40 and 250 measurements during the five-year observations (Taris et al. 2018). It is possible to read about the telescope itself, CCDs, astrometric and photometric measurements, Radial Velocity Spectrometer (for spectroscopy), scanning process, some technical parts, etc. in the paper (Prusti 2012).

2. GAIA DATA RELEASE 1

Since 14 September 2016, the first release of the Gaia catalogue (DR1) is available, as a first step of the future Gaia celestial reference frame (Gaia CRF). The optical Gaia positions of the sources will be materialization of that reference frame. Gaia CRF will be linked to the ICRF which is based on the VLBI radio observations of extragalactic sources. The Gaia is collecting data for more than one billion sources brighter than G magnitude 20.7 (the white-light photometry band of Gaia) to determine high accurate positions, proper motions and parallaxes; Gaia catalogue is main goal of that ESA mission. The number of QSOs is about 500 000. The DR1 is based on the first observational period of about 14 months, and in that catalogue there are near two million stars using Tycho-Gaia solution.

In the DR1, only data were published about flux time-series variability detection for Cepheids and RR Lyrae, but not for AGN and particularly QSOs.

In the paper (Carrasco et al. 2016) is described the principles of the photometric calibrations of the G band.

At the beginning of the XI Bulgarian-Serbian Astronomical Conference, during registration and abstract phase of conference, it was published only DR1, but just before that conference (14 - 18 May 2018 in Belogradchik, Bulgaria) the DR2 was available (at April 2018). Because of it, in this paper there is more information about DR2 than about DR1. Other information about DR1 could be found in the paper (Taris et al. 2018).

The final Gaia solution (catalogue) will be consisting of: astrometric, photometric, and radial-velocity data, variable-star and non-single-star results, object classifications with multiple astrophysical parameters for stars, QSOs, galaxies, and unresolved binaries, exo-planets, epochs and transits for all objects.

3. GAIA DATA RELEASE 2

The second release of the Gaia catalogue - DR2 was released on 25 April 2018. It is based on 21 months or 1.75 yr, 640 days with some interruptions (period 22 August 2014 – 23 May 2016) of Gaia operational phase, even the start of astronomical observations was in July 2014. The first month of Gaia observations was not included in DR2 solution because of not enough quality of these data.

That catalogue contains results for 1.693 billion sources in the magnitude range 3 to 21. For 1.332 billion sources there are the five astrometric parameters: positions, proper motions, and parallaxes. The reference epoch is J2015.5 = JD 2457206.375 TCB = 2 July 2015 at 21:00:00 TCB, and it is about half-way through the observational period used in the DR2 solution (that epoch is 0.5 yr later than the DR1 one). Because of it, there are some differences in positional data between DR1 and DR2 releases. The reference epoch was chosen to get minimal correlations between the positions and proper motions. The positions and proper motions refer to the ICRS.

Also, there are the approximate positions for an additional 0.361 billion mostly faint objects (Lindegren et al. 2018). The DR2 is publicly available in the online Gaia Archive (https://archives.esac.esa.int/gaia).

The catalogue DR2 is independent because it does not include any other astrometric data (Hipparcos or Tycho ones), in contrast of the DR1 which is the Tycho-Gaia astrometric solution (Lindegren et al. 2016). All sources are reduced as single stars (as in DR1 solution) and the results are presented by the five astrometric parameters.

The results of some binary stars refer to the photocentre (in case of unresolved binaries) or to either component (for resolved ones).

The main steps of astrometric solution (models, algorithms, etc.) are described in Lindegren et al. (2012), but there are some additions since 2012.

In DR2, there is the color information for most of the sources using the photometric processing of data via the blue BP and red RP photometers (van Leeuwen et al. 2017).

The astrometric calibration parameters of the CCDs were determined using an astrometric solution for 16 million selected objects (about 1% of the input data), and via QSOs (as extragalactic sources) the primary solution was linked to the ICRS. After that, the astrometric parameters of other sources were calculated (Lindegren et al. 2018).

Figure 1: The Gaia spacecraft or space telescope (artist's impression).

For bright sources (G < 14 mag), the median uncertainty is near 0.04 mas in parallax and position (at J2015.5 – reference epoch of DR2). For G=17 mag, it is 0.1 mas, and 0.7 mas in the case of G=20 mag. About the proper motions, there are 0.05 mas, 0.2 mas, and 1.2 mas, respectively.

The optical Gaia DR2 CRF is aligned with ICRS. It is non-rotating to within 0.15 mas per year with respect to the QSOs (Lindegren et al. 2018).

The systematic in the parallaxes are less than 0.1 mas (depending on positions, magnitude, and color).

A consistent theory of relativistic astronomical reference systems was used during the processing of Gaia data (Soffel et al. 2003). The Barycentric Celestial Reference System – BCRS is the primary coordinate system: the origin is at the solar system barycentre, the axes are aligned with the ICRS, the barycentric coordinate time – TCB is the time-like coordinate of the BCRS.

In DR2 solution is not included non-linear motions (as it is in the case of binary stars, other perturbations, etc.), but the future Gaia releases will include these motions (Lindegren et al. 2018). In DR2, only the uniform space motion (of the source) relative to the solar system barycentre was considered.

So, the Gaia-CRF2 is the celestial reference frame of Gaia DR2; it is aligned with ICRS and non-rotating with respect to the distant objects (QSOs). There are 2843 sources – mostly QSOs (as the optical counterparts of VLBI radio sources) in a prototype version of ICRF3. The prototype ICRF3 catalogue (unpublished, 2017) contains 4262 sources with accurate VLBI radio positions. For ICRF3, the IAU Working Group “Third Realisation of ICRF” is responsible.

As it was a case of DR1, in DR2 there are only photometric data as time-series for Cepheids and RR Lyrae (the objects with unstable flux), but not for QSOs.

About the Gaia DR3 (next solution), it is planed for the first half of 2021. That improved astrometry and photometry solution will be consisting of: mean Vr velocities for stars without detected variability, object classification and astrophysical parameters (BP/RP and RVS spectra for spectroscopically objects), with the epoch photometry it will be variable-star classifications, Solar-system results (preliminary orbital solutions, individual epoch observations), non-single star catalogues, etc.

Figure 2: The 1.4 m telescope (left) with new dome (right) at ASV.

4. SERBIAN-BULGARIAN ACIVITIES IN LINE WITH GAIA MISSION

Our observational activities and investigation about the Gaia mission and Serbian-Bulgarian cooperation in line with Gaia tasks stated after the installation of the fist instrument, the 60 cm telescope (Fig. 4), at the Serbian new site (see

Fig. 5), the Astronomical Station Vidojevica – ASV. The ASV belongs to Astronomical Observatory in Belgrade – AOB (http://vidojevica.aob.rs), and it is close to Prokuplje city (southern Serbia, see Fig. 5). That installation was done in 2011. Since mid-2016, there is another ASV instrument via Belissima project (see http://belissima.aob.rs), 1.4 m ASV telescope (see Fig. 2, the telescope itself and new dome).

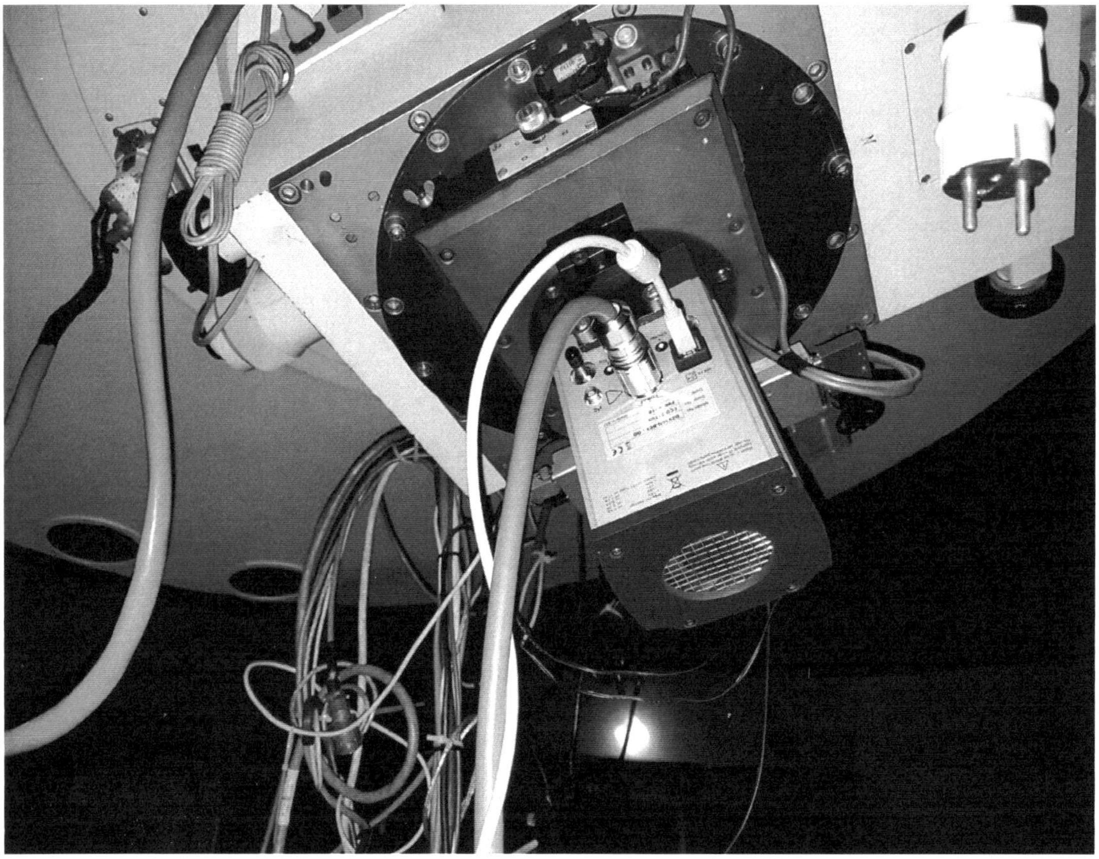

Figure 3: The 2 m telescope (NAO Rozhen) with new CCD camera Andor iKon-L.

At two Bulgarian sites (see Fig.5), Belogradchik and Rozhen,, there are another 4 instruments of our interest. In Belogradchik, it is the 60 cm telescope. And at the Rozhen Observatory there are three instruments: the Schmidt – camera 50/70 cm, 2 m (see Fig. 3) and 60 cm telescopes. Using these 6 telescopes (Damljanović, Vince and Boeva 2014; Damljanović et al. 2018; Damljanović, Taris and Andrei 2018; Taris et al. 2018) we established astronomical cooperation "Serbian-Bulgarian mini-network telescopes" in 2013 (head – G. Damljanović). The 2 m Rozhen telescope and its new CCD camera Andor iKon-L (Fig. 3) have new features: 2048x2048 pixels, the pixel size is 13.5x13.5 mkm, 0.176 arcsec per pixel, FoV=6.0x6.0 arcmin. That camera was put on the instrument at April 2018.

Our activities about the Gaia tasks are in accordance with the bilateral Serbian-Bulgarian joint research project "Observations of ICRF radio-sources visible in optical domain" during three years period 2014-2016, and new one "Study of ICRF radio-sources and fast variable astronomical objects" for the period 2017-2019; the head is G. Damljanović. Both projects are in the framework between the Serbian Academy of Sciences and Arts – SASA (or SANU on Serbian language) and Bulgarian Academy of Sciences - BAS.

Figure 4: The 60 cm telescope at ASV.

About the future Gaia reference frame, the QSOs optical flux variations were investigated using our original observations of QSOs (in the period 2013-2015) by mostly the TAROT telescopes, 60 cm ASV and other instruments (Taris et al. 2018). The mentioned Gaia reference frame will be materialized by the optical positions of the objects. And it will be linked to the ICRF (based on the VLBI radio positions of mostly QSOs). First at all, it is necessary to investigate flux variability of QSOs because the unstable flux can indicates some changes in the source structure which is of importance for position of the target photocentre (it means, for position of the object and results of Gaia astrometry). Also, it is possible the evolution in time of these centers. It could produce consequences for mentioned link. Some magnitude variations of 47 objects (Bourda et al. 2008, Bourda et al. 2011), mostly QSOs, which are suitable for the link are presented in our papers (Taris, Damljanović et al. 2018). These papers are based on the 60 cm ASV data (and other telescopes ones), and reports on some implications for the Gaia catalogue.

The sites ASV, AO Belogradchik and Rozhen Observatory (or National Astronomical Observatory – NAO of BAS) are presented on the picture of Serbian-Bulgarian region (see Fig. 5).

Figure 5: The ASV, Belogradchik and Rozhen astronomical sites.

The Gaia astrometry is the main part of our investigation in accordance with Serbian-Bulgarian cooperation, but it is the Gaia Alerts, also. During three years period, October 2014 – October 2017, we observed about 45 objects (near 1650 CCD images) of the Gaia Alerts or Gaia-Follow-Up Network for Transients

Objects (Gaia–FUN-TO) using mentioned 6 telescopes. It is about 15 objects per year (near 550 CCD images). Some results were presented at few conferences, and there are few published papers (Campbell et al. 2015, Damljanović et al. 2014). We are doing with the Johnson BV and Cousins RcIc filters. Mostly, the 60 cm ASV telescope was used. Opposite, no data from 60 cm telescope of NAO Rozhen during last few years, because it was under reconstruction until the summer this year. We hope, it will be some observations in the domain of Gaia Alerts, soon.

At Serbian-Bulgarian sites, the seeing is between 1.0 and 3.5 arcsec. The mean value at ASV is 1.2 arcsec, and it could be 0.7 arcsec at Rozhen and ASV. It is possible to observe the objects until about 20 mag in V-filter by using the 2 m of NAO Rozhen and 1.4 m at ASV telescopes (if the exp. time is near 5 min), and until 19 mag using smaller instruments.

5. CONCLUSIONS

The second Gaia solution contains a vastly increased number of sources in comparison with Gaia DR1. These sources have full astrometric data (proper motions and parallaxes, also). The data of bright sources (it means, G is until 12 mag) were included already in DR1, but the results in DR2 are generally more accurate. The results of DR2 are fully independent of the Hipparcos and Tycho catalogues. The Gaia CRF2 (reference frame) is defined by Gaia data of QSOs; also, the optical counterparts of VLBI objects in a prototype solution of the ICRF3.

Random and systematic errors are still higher than can be expected for the final solution, because the astrometric results in Gaia DR2 are based on about 2 years of observations. Also, calibrations are very preliminary, and there are possibilities for improvements (Lindegren et al. 2018). It means, there are some investigations to do for the next DR3 solution. Other job to do, in DR2 all objects beyond the solar system are treated as point sources and it is necessary to improve in the reduction model. The next limitation of DR2 is about unresolved binaries where the photocentre was calculated instead of each separate star of the system. The systematic errors are mainly based on the analysis of QSOs data, and it is a part for improving Gaia DR3.

About our activities in line with Gaia: we established the cooperation "Serbian-Bulgarian mini-network telescopes" in 2013, together SANU-BAS projects, we did some investigation in accordance with Gaia astrometry (Taris et al. 2018), Gaia Alerts or Gaia-FUN-TO (Campbell et al. 2015, Damljanović et al. 2014), etc. The old SANU-BAS project was "Observations of ICRF radio-sources visible in optical domain" during three years period 2014-2016, and new one is "Study of ICRF radio-sources and fast variable astronomical objects" for the period 2017-2019. During three years period (October 2014 – October 2017) we observed about 45 Gaia Alerts Objects or near 1650 CCD images. It is about 15 objects per year or near 550 CCD images.

The Gaia is continue its five–year period of observations (that period is going to prolong), and we continue our observational activities and investigation in line with that very important ESA mission.

Acknowledgements

I acknowledge the observing grant support from the Institute of Astronomy and Rozhen National Astronomical Observatory, Bulgarian Academy of Sciences, through the bilateral joint research project "Study of ICRF radio-sources and fast variable astronomical objects" (for the period 2017-2019, head – G. Damljanović). This work is a part of the project No. 176011 "Dynamics and kinematics of celestial bodies and systems" supported by the Ministry of Education, Science and Technological Development of the Republic of Serbia.

References

Bourda, G., Charlot, P., Le Campion, J.: 2008, *Astronomy and Astrophysics,* **490**, 403.
Bourda G., Collioud A., Charlot P., et al: 2011, *Astronomy and Astrophysics,* **526**, A102.
Campbell, H. C., et al: 2015, *MNRAS,* 452, 1960.
Carrasco J., Evans, D., Montegriffo, P., et al: 2016, *Astronomy and Astrophysics,* **595**, A7.
Damljanović, G., Vince, O., Boeva, S.: 2014, *Serb. Astron. J.,* **188**, 85-93.
Damljanović, G., Taris, F., Andrei, A.: 2018, *Astronomy and Astrophysics in the Gaia sky Proc., IAU Sym. No.,* **330** (2017), A. Recio-Blanco, P. de Laverny, A.G.A. Brown and T. Prusti, eds., 88-89.
Damljanović, G., Latev, G., Boeva, S., Vince, O., Bachev, R., Jovanović, M. D., Cvetković, Z., Pavlović, R. : 2018, *Publ. Astron. Obs. Belgrade,* **98**, 277-280.
ESA: 1997, *The Hipparcos and Tycho Catalogues, ESA SP*-1200.
Lindegren, L., Lammers, U., Hobbs, D., et al.: 2012, *Astronomy and Astrophysics,* **538**, A78.
Lindegren, L., Lammers, U., Bastian, U., et al.: 2016, *Astronomy and Astrophysics,* **595**, A4.
Lindegren, L., Hernandez, J., Bombrun, A., et al.: 2018, *Astronomy and Astrophysics,* **616**, A2.
Prusti, T.: 2012, *Astron. Nachr.,* 333, **5/6**, 453–459.
Soffel, M., Klioner, S. A., Petit, G., et al: 2003, *AJ,* **126**, 2687.
Taris, F., Damljanović, G., Andrei, A., Souchay, J., Klotz, A., Vachier, F.: 2018, *Astronomy and Astrophysics,* **611**, A52.
van Leeuwen, F.: 2007, *Hipparcos, the New Reduction of the Raw Data,* Astrophysics and Space Science Library, 350.
van Leeuwen, F., Evans, D. W., De Angeli, F., et al.: 2017, *Astronomy and Astrophysics,* **599**, A32.

Proceedings of the XI Bulgarian-Serbian Astronomical Conference (XI BSAC) Belogradchik, Bulgaria, May 14-18, 2018
Editors: Milcho K. Tsvetkov, Milan S. Dimitrijević and Momchil Dechev
Publ. Astron. Soc. "Rudjer Bošković" No 18, 2018, 23-36

BBN COSMOLOGICAL CONSTRAINTS ON PHYSICS BEYOND THE STANDARD MODEL

DANIELA KIRILOVA

Institute of Astronomy and NAO, Bulgarian Academy of Sciences
Blvd. Tsarigradsko Shosse 72, Sofia, 1784, Bulgaria
E-mail: dani@astro.bas.bg

Abstract. Big Bang Nucleosynthesis is the most reliable probe of the physical conditions of the early Universe during the cosmological nucleosynthesis epoch. Therefore, BBN is traditionally used as a cosmological test of physics beyond the Standard Model. We review several non-standard BBN models and the obtained cosmological constraints on the baryonic density, the number of neutrino families, neutrino mixing and neutrino masses, additional particles and interactions, in particular on chiral tensor particles coupling.

1. INTRODUCTION

Astrophysical and cosmological observations data requires the existence of beyond Standard Model (BSM) physics. For example, the contemporary cosmological model – Lambda Cold Dark Matter Model (LCDM) contains components called Dark Energy (DE) and Dark Matter (DM) with unknown nature, which constitute 95% of the Universe matter! The necessity of early inflationary period and the generation of the observed baryon asymmetry of the Universe also require BSM ingredients for their realization (to propose the DM, DE, inflaton candidates). The other possibility is to change the theoretical basis of the standard cosmological model – to propose an alternative gravitational theory, etc.

BSM physics in the neutrino sector – neutrino oscillations, is already experimentally firmly established. Neutrino oscillations experiments challenged the standard model assumptions of zero neutrino masses, no mixing, 3 light neutrino families, lepton number conservation, equilibrium Fermi Dirac distribution.

This review is mainly dedicated to the Big Bang Nucleosynthesis (BBN) constraints on BSM neutrino physics. In the next section I discuss contemporary status of BBN as the deepest reliable early Universe probe and BSM physics test.

The third section discusses BBN as Universe baryometer, which constrains the matter content of the Universe, points to the existence of hidden baryons, nonbaryonic dark matter and to the necessity of baryogenesis, in case the observed locally matter-antimatter asymmetry of the Universe is its global characteristic. Then BSM neutrinos and BBN constraints on inert neutrino, number of neutrino families and lepton asymmetry (L) in the neutrino sector are briefly discussed. The fourth section discusses BBN with neutrino oscillations and with L, neutrino oscillations - lepton asymmetry interplay and the possible solution to the dark radiation (DR) problem. The fifth section discusses BSM with chiral tensor particles and their cosmological influence and BBN constraints.

2. BBN - THE DEEPEST RELIABLE EARLY UNIVERSE PROBE

First ideas about BBN, that appeared in the period 1946-1948, belong to George Gamow. With his collaborators he developed BBN theory and predicted the existence of Cosmic Microwave Background (CMB). BBN describes the production of the light elements D, He-4, He-3 and Li-7 and some tiny traces of heavier nuclei Be-9, B-10, B-11 up to CNO isotopes during the early hot and dense radiation dominated stage (RD) of the Universe, corresponding to the period from the first second to the first 20 minutes. For recent reviews see Cyburt et al. (2016), Partignani et al. (2017), Pitrou et al (2018).

Contemporary BBN is theoretically well established - based on well-understood Standard Models physics, namely General Relativity, LCDM cosmology, Standard model of nuclear and particle physics. The primordially produced abundances are usually parameterized by the baryon-to-photon ratio $\eta=n_b/n_\gamma$, relativistic energy density (effective number of light neutrino families (N_{eff}) and neutron life time τ_n. All these have been already measured, namely, η was determined independently by the analysis of the anisotropies of the CMB (Ade et al., 2016):

$$\eta_{CMB} = (6.11\pm0.04).\ 10^{-10},\ 68\%\ \text{CL}.$$

LEP experiments at CERN determined:

$$N_{eff} = 2.984\pm0.008.$$

τ_n has been revised recently (Serebrov et al. 2017):

$$\tau_n = 879.5\pm0.8\ \text{s}.$$

Thus, contemporary BBN is a parameter free theory. The predicted abundances depend on the well measured cross sections of nuclear processes, which have been continuously updated and on the well measured neutron lifetime. Over 400 nuclear reactions are considered. Modern analyses of nuclear rates for BBN have been provided (see for example NACRE compilation of Angulo et al. 1999 and the updated NACRE-II compilation Xu et al. 2013). During the last decades more and more precise BBN codes have been invented, like PArthENoPE (Pisanti et al., 2008; Consiglio et al., 2017), AlterBBN (Arbey, 2012), PRIMAT (Pitrou et al., 2018), following the first BBN codes of Wagoner

(1967) and Kawano (1992). BBN calculations of the light elements by several different groups are in good agreement. BBN predicted abundances then are compared with the precise observational data on light elements abundances.

The primordially produced abundances are obtained from observational data in systems least contaminated by stellar evolution, after that the account for galactic chemical evolution is made. Namely, D is measured in high redshift z low-metallicity (low Z) H-rich clouds absorbing light from background QSA. He-4 is measured from emission lines in highly ionized H gas of the most metal-poor blue compact galaxies (extragalactic H II regions), after which a regression to zero Z is provided. Li is measured in Population II (metal-poor) stars in the spheroid of our Galaxy, which have $Z<1/10\ 000\ Z_{Sun}$.

The last few years the precision of observational data on primordial abundances has been drastically improved. New observations of QAS have considerably improved the observational determination of D and lowered its uncertainty. The latest observational value for primordially produced D is (Cooke et al., 2017):

$$D/H=(2.527\pm0.03)\ 10^{-5}.$$

Recently, the emissivities of He-4 were updated and a new infra-red line was added (Izotov et al 2014), which led to more precise abundance determination in good agreement with BBN predicted one. He-4 primordial mass fraction is determined with 10^{-4} accuracy (Pitrou et al. (2018)):

$$Y_p=0{,}24709\pm 0{,}00017\ .$$

The primordial abundance of Li is given by (Sbordone et al. (2010)):

$$Li/H=(1.58\pm0.31)\ 10^{-10}.$$

It is by a factor of 3 less than the BBN predicted value. This Li-7 problem is considered as an indication for BSM physics. Some of the proposed BSM solutions include modified nuclear rates, new resonant interactions, new particles decaying before or during BBN (Dolgov&Kirilova, 1986), variation of fundamental constants (Dmitriev et al. 2004, Coc et al. 2007, 2012), etc.

It is remarkable that, the predicted abundances (except Li-7) are in good overall agreement with those inferred from observational data for $\Omega_B \sim 0.05$, and also with those "measured" by CMB. This allows the use of BBN as the earliest precision probe of physical conditions in the early Universe and also as the earliest test for BSM physics, corresponding to the BBN energy diapason (1 MeV- 10 KeV). Thus, BBN is used as precise Universe baryometer during BBN epoch, as the best speedometer at RD stage, as the most exact Universe leptometer (as will be discussed in the next sections), etc.

BBN is able to constrain physics beyond standard model thanks to the good concordance between BBN theory and observational data and thanks to the fact that BBN depend on all known interactions. Hence, it constrains the modification of those. Some of the often considered BBN constraints include: constraints on additional light (relativistic during BBN, i.e. m< MeV) particles species (generations) effecting radiation density (and correspondingly the Universe expansion rate H), pre-BBN nucleon kinetics or BBN itself; constraints on additional interactions or processes relevant at BBN epoch (decays of heavy particles, neutrino oscillations); constraints on the possible departure from equilibrium distributions of particle densities of nucleons and leptons (caused by neutrino oscillations, lepton asymmetry, inhomogeneous distribution of baryons, etc.); constraints on SUSY, string models, extradimensional models, etc.

3. BBN - BARYOMETER, SPEEDOMETER AND LEPTOMETER AT RADIATION DOMINATED STAGE

BBN was until recently considered the *unique universe baryometer*. The baryon density for which BBN predicted light element abundances values compatible with the ones obtained from observations is:

$5.8 \times 10^{-10} < \eta_{BBN} < 6.6 \times 10^{-10}$ 95%

corresponding to

$$0.021 \leq \Omega_b h^2 \leq 0.024 (95\% C.L.)$$

where

$$\Omega_b h^2 = 3.65 \times 10^7 \eta, \quad \Omega_b = \frac{\rho_b}{\rho_c}, \quad \rho_c = \frac{3H^2}{8\pi G_N}$$

Deuterium is the most sensitive element to the baryon density among the light elements produced primordially and is used as a powerful baryometer and a test of the concordance between BBN and CMB baryon measurements. The baryon density determined on the basis of D measurements and BBN is (Pitrou et al., 2018):

$$\Omega_b h^2 = 0.0219 \pm 0.00025 (95\% C.L.)$$

These results are in excellent agreement with the determinations of the baryon density by the latest CMB anisotropy measurements (Planck 2016) corresponding to a much later epoch of the Universe evolution (z~1000). Thus, BBN remains the best baryometer at the RD stage. Assuming that the baryon density has not changed between BBN and CMB epochs, the analysis of BBN and CMB together provides more precise determination:

$$\Omega_b h^2 = 0.0215 \pm 0.00014 (95\% C.L.)$$

The baryon density is much bigger than the luminous matter density, which means that most of the baryons are optically dark. It has been observationally found that half of the dark baryons are in the space between galaxies. Namely, in the spectra of distant quasars the absorption lines of ordinary baryonic matter were found (Danforth & Shull, 2008). The other half of dark baryons is supposed to be hidden in MACHOS, black holes, etc.

The baryon density is 0.049 of the total density. Besides, combined results of Hubble Space Telescope and CMB measurements and galaxy clusters data point to the existence of gravitating matter, constituting 26.8% of the total density, i.e. much bigger than the baryon density. BBN points to the existence of nonbaryonic dark matter, requiring BSM physics for the explanation of DM origin. The rest of the Universe density- 68.3% is in the form of DE, causing the observed by SNIa accelerated expansion of the Universe.

It is assumed that BSM physics must be invoked to answer the questions: Why baryonic matter is such a small fraction? What is the nature of nonbaryonic matter? Where are the antibaryons? How and when the net baryon number was generated? Is the asymmetry local or global?

BBN measured baryon density value is used as an observational constraint for baryogenesis models, aiming to explain the generation of the excess of baryons over antibaryons in the Universe. Standard cosmology predicts equal quantities at the hot stage and now the expected relic density is: $\beta \sim 10^{-18}$, which is by many orders of magnitude smaller than the observed value $\beta_{obs} \sim \eta$. The explanation of the measured value of the baryon asymmetry requires BSM physics. Most often baryogenesis models require B number violation and CP-violation BSM nonequilibrium processes.

Cosmic ray data from search of antiprotons, positrons and antinuclei indicate that there is no significant quantity of antimatter objects within a radius 1 Mpc. Gama ray data point that no significant amounts of antimatter exist up to galaxy cluster scales ~ 10 -20 Mpc. However, above that scale both theory and observations allow astronomically significant quantities antimatter. In case the baryon asymmetry is local, it is necessary to find acceptable separation mechanisms between the domains of matter and antimatter. Such kind of mechanisms also include BSM physics. See for example the inhomogeneous baryogenesis models of Dolgov&Kirilova (1989), Kirilova&Chizhov (2000), Kirilova(2002), Dolgov et al. (2008).

Besides being a unique baryometer at the RD stage, BBN is an *excellent speedometer*. He-4 is very sensitive element to the expansion rate of the Universe at the BBN epoch, which is usually parameterized by the effective number of the light neutrino types N_{eff} (Shvartsman 1969). Different types of BSM physics predict extra relativistic component ΔN_{eff} due to sterile neutrino, neutrino oscillations, lepton asymmetry, neutrino decays, nonstandard thermal history, etc

A maximum likelihood analysis by Cyburt et al. (2016), provided the following contemporary BBN constraints on N_{eff} and the baryon-to-photon ratio:

$$2.3 < N_{eff} < 3.4$$

$$5.6 < \eta < 6.6$$

Most recent stringent BBN constraint was obtained by Pitrou et al. (2018):

$$N_{eff} = 2.88^{+0.27}_{-0.27} \quad (95\%).$$

The CMB constraint (Planck Collaboration 2015) reads:

$$N_{eff} = 3.13^{+0.31}_{-0.31} \quad (95\%)$$

Now, as a result of the improved determination of D and the contemporary precision BBN, D provides more stringent constraints on N_{eff} than He-4. Using CMB plus D plus He-4 data allows a considerable reduction of the error (Cyburt, 2016):

$$N_{eff} = 2.88^{+0.16}_{-0.16} \quad (95\% \text{ Planck+D+He-4})$$

The most stringent constraint of CMB plus BBN now is:

$$N_{eff} = 3.01^{+0.15}_{-0.15} \quad (95\% \text{ Planck +BBN})$$

Thus, BBN and CMB constraint strongly all BSM physics introducing additional light species, as for example supersymmetric scenarios (lightest particle neutralino or gravitino), string theory, large dimensions theories, GUT with sterile (right handed) neutrinos, decaying particles, SUSY metastable particles, etc.

BBN is known also to be an exact *Universe leptometer*. Dynamical and kinetic effect of lepton asymmetry L on BBN lead to the BBN bound $|L|<0.1$. BBN bounds on L are changed in case of neutrino oscillations. Stringent BBN constraints on L in case of electron-sterile oscillations exist, namely L as small as 10^{-8} may be felt by BBN with oscillations (Kirilova 2012). In more detail BBN as a leptometer is discussed in 4.2.

4. BBN WITH NEUTRINO OSCILLATIONS AND LEPTON ASYMMETRY

4.1. BBN with Neutrino Oscillations

BSM physics of BBN with vacuum neutrino oscillations *proceeding before electron neutrino decoupling* was first studied by Dolgov (1981). In particular, it was found that active-sterile oscilltions may excite into equilibrium the sterile neutrino state, thus increasing the expansion rate of the Universe and influencing BBN. Cosmological effects of such fast neutrino oscillations in medium were first studied by Barbieri&Dolgov (1990, 1991), where besides the dynamical effect of neutrino oscillations also the depletion of the electron neutrinos due to oscillations was found. Cosmological constraints on neutrino oscillation parameters were obtained, which excluded large mixing angle (LMA) sterile solution to the solar neutrino problem.

BBN with non-equilibrium active-sterile neutrino oscillations in vacuum, *proceeding after electron neutrino decoupling*, was first studied by Kirilova (1988) and BBN with non-equilibrium matter active-sterile neutrino oscillations were first studied by Kirilova&Chizhov (1996,1997). It was found that active-sterile oscillations proceeding after decoupling may strongly distort neutrino energy spectrum from the equilibrium Fermi-Dirac form and enhance lepton asymmetry in the neutrino sector. BBN was determined to be a sensitive probe to additional species and to distortions in the neutrino distribution. First BBN constraints on neutrino oscillation parameters were obtained in these original papers accounting for the dynamical effect of neutrino oscillations and for the spectrum distortion effect and lepton asymmetry generation.

In several works (Kirilova&Chizhov, 1997, 1998a, 1998b, 2000; Kirilova, 2004, 2007, 2012; Kirilova&Panayotova, 2006) we have studied numerically the evolution of neutrino ensembles, evolution of L, and the evolution of nucleons during pre BBN epoch. Numerical analysis of the produced primordial helium-4 dependence on oscillation parameters (neutrino mixing and squared mass differences), lepton asymmetry value and the population of the sterile neutrino state Y_p (δm^2,θ, L, δN_s) has been provided for 10^{-10}<L<0.01, for the full parameter range of parameters of the oscillations model and for the BBN temperature range, namely

$$\delta m^2 \leq 10^{-7} eV^2 \qquad \textit{all mixing angles } \theta \quad 0 \leq \delta N_s \leq 1$$
$$2\ MeV \geq T \geq 0.3\ MeV$$

Precise BBN constraints on oscillation parameters were determined, accounting for all cosmological effects of neutrino oscillations discussed in this BSM of BBN.

The fit to these BBN constraints, corresponding to 3% He-4 overproduction and initially empty sterile neutrino state, reads:

$$\delta m^2 \left(\sin^2 2\theta\right)^4 \leq 1.5\times 10^{-9} eV^2 \quad \delta m^2 > 0$$
$$\delta m^2 < 8.2\times 10^{-10} eV^2 \quad \text{large } \theta,\ \delta m^2 < 0$$

These BBN constraints were by 4 orders of magnitude more stringent than experimental ones. They excluded LOW active-sterile solutions (1990, 1999) years before experimental results.

BSMs of BBN with neutrino oscillations excluded LMA and LOW active-sterile solutions years before experimental results of neutrino oscillations experiments managed to exclude the active-sterile solutions to the neutrino anomalies.

BSMs of BBN with neutrino oscillations and non-zero ν_s population were considered as well (Kirilova 2004, 2007 ; Kirilova&Panayotova 2006). It was found that additional ν_s population change non-trivially the BBN bounds on oscillation parameters due to the interplay between the dynamical and kinetic effects of non-zero initial population of ν_s (partially filled) on BBN. It may strengthen or relax BBN constraints. In case the dynamical effect dominates,

He-4 overproduction is enhanced and BBN constraints strengthen.

In case the kinetic effect dominates He-4 overproduction decreases with δN_s increase and BBN constraints relax.

4.2. BBN with Neutrino Oscillations and Lepton Asymmetry

BSMs of degenerate BBN have been considered since the original paper of Wagoner et al. (1967). Lepton asymmetry dynamical effect – the increase of the radiation energy density

$$L = \sum_i \frac{1}{12\zeta(3)} \frac{T_{\nu_i}^{\ 3}}{T_\gamma^{\ 3}} (\xi_{\nu_i}^3 + \pi^2 \xi_{\nu_i})$$

$$\Delta N_{\text{eff}} = 15/7((\xi/\pi)^4 + 2(\xi/\pi)^2)$$

has been studied. This leads to faster expansion $H = (8/3\pi G\rho)^{1/2}$, delay of matter/radiation equality epoch, influence BBN, CMB, evolution of perturbations i.e. LSS. For a review see Lesgourgues&Pastor (1999).

Besides, L in the electron neutrino sector has direct effect on neutron-proton kinetics in pre-BBN epoch. Therefore, BBN provides stringent constraints on L_e. In case of BBN with neutrino oscillations degeneracies in different sectors equilibrate due to oscillations before BBN (Dolgov et al. 2002). Then, the following BBN constraint on neutrino degeneracy parameter holds:

$$|\xi_\nu| < 0.1$$

More refined BBN constraints on L were discussed in the following works, see for example Miele et al. (2011), Steigman (2012), Mangano et al. (2013), which accounted for flavor oscillations and ν decoupling, Recent improvement on D and He-4 measurement allowed more stringent BBN constraints:

For comparison CMB $|\xi_\nu| < 0.016 (68\% CL)$ $\qquad L < 0.01$

Different *indirect kinetic effec* $|\xi_\nu| < 0.06 (68\% CL)$ ı neutrino evolution, its number density, spectrum distribution distortion, oscillations pattern and hence on n/p kinetics and BBN was found and studied (Kirilova&Chizhov 1998, 2000; Kirilova, 2011, 2012). It was found that BBN with electron-sterile oscillations feels L<< 0.01.

An interesting interplay between ν oscillations and lepton asymmetry L was found possible in studies of BBN with active-sterile neutrino oscillations and L: (i) *Neutrino active-sterile oscillations change neutrino-antineutrino asymmetry of the medium:* they can suppress pre-existing asymmetry (Barbieri &Dolgov1991; Enqvist et al. 1992) or enhance L in MSW resonant active-sterile oscillations for $\delta m^2 \sin^4 2\theta < 10^{-7} eV^2$ in the collisionless case (Kirilova&Chizhov 1996) and $\delta m^2 > 10^{-5} eV^2$ in oscillations dominated by collisions (Foot et al. 1996).

For non-equilibrium neutrino oscillations between ν_e and ν_s, effective after ν_e decoupling, i.e. $\delta m^2 \sin^4 2\theta < 10^{-7} eV^2$, the region of parameter space for which large generation of L is possible was determined (Kirilova, 2012) :

$$|\delta m^2| \sin^4 2\theta \leq 10^{-9.5} \mathrm{eV}^2$$

(ii)Lepton asymmetry effects neutrino oscillations: it may suppress oscillations (Foot&Volkas 1995; Kirilova&Chizhov 1998) *or enhance oscillations* (Kirilova&Chizhov 1998). On the basis of numerical analyses relations have been derived between L and neutrino oscillations parameters, corresponding to different L influence (Kirilova, 2012). Recently an update of this analysis was provided (Kirilova, 2018) allowing to derive the following relation connecting neutrino squared mass difference and L value necessary to inhibit neutrino oscillations.

$$L > (0.01\delta m^2 / eV^2)^{3/5}$$

In BBN with neutrino oscillations spectrum distortion and L generation lead to different nucleon kinetics, and modified BBN element production. Hence, modified BBN constraints on oscillation parameters in presence of L were obtained (Kirilova&Chizhov 1998b, 2000; Kirilova, 2012). The account of the neutrino-antineutrino asymmetry growth caused by resonant oscillations leads to relaxation of the BBN constraints for small mixings.

In case of small relic L, namely $10^{-11} < L < 0.01$, due to its indirect kinetic effect L change primordial production of He-4 by enhancing or suppressing oscillations depending on the interplay between its value and the neutrino oscillation parameters. Hence, relic L may strengthen, relax or eliminate standard BBN bounds on oscillation parameters. This modified BBN model with late active sterile neutrino oscillations is very sensitive leptometer – it feels L as small as 10^{-8} (Kirilova, 2012).

4.3. Dark Radiation Problem and Its BSM Solution

Dark radiation (DR), i.e. the presence of non-zero ΔN_{eff} has been predicted by different BSM theories. DR is of particular interest today in view of experimental indications from neutrino oscillations short baseline (SBL) experiments for ν_s with mass in the eV range.

Anomalous results of neutrino oscillations SBL experiments data including reactor experiments+LSND+MiniBooNe+Gallium (GALLEX, SAGE) suggested the existence of light ν_s with 1.3 eV mass, with mixing with flavor neutrinos ν_a $\sin^2\theta_{14}$ in the range [0.01-0.03] (Giunti, 2017; Gariazzo et al. 2018; Capozzi et al, 2017).

However, stringent cosmological constraints on additional dark radiation ΔN_{eff} exist, as discussed in the previous section. It is known that eV sterile neutrino is brought into equilibrium in the early Universe due to fast oscillations between flavor and sterile neutrinos $\nu_a \leftrightarrow \nu_s$. This influences CMB and BBN through increasing the radiation density and the Universe expansion rate. Additional eV neutrinos lead to overproduction of He-4, i.e. BBN constraints additional light neutrinos (Dolgov 1981, Barbieri&Dolgov 1990). The same holds for CMB (see the contemporary BBN constraints on ΔN_{eff} discussed above). Besides, recently the following constraints have been obtained on the basis of Lyman Alpha forest BOSS data, CMB data from Planck, ACT, SPT, WMAP polarization (Rossi Yeche et al. 2015):

$$N_{eff} = 2.911^{+021.}_{-0.22} \quad 95\%\ \text{C.L}$$
$$\Sigma m_\nu < 0.15\ \text{eV}$$

See also similar stringent constraints in case data from baryon acoustic oscillations are added, discussed by Sasankan et al. (2017). Thus, cosmology constraints severely the thermalized during BBN light sterile neutrinos.

Different BSM deviations from standard LCDM have been invented to solve the DR problem, including additional radiation, change in matter density, decaying particles during BBN, etc. We proposed the interplay between L and neutrino oscillations as a solution. Namely, L suppresses $\nu_s \leftrightarrow \nu_e$ oscillations, thus preventing ν_s thermalization, and avoiding cosmological constraints on eV sterile neutrinos (Kirilova 2012, 2013). See also the works by Mirizzi et al. (2012), Hannestad et al. (2012), dedicated to this type of DR problem solution.

Recently we updated the analysis of the interplay between $\nu_s \leftrightarrow \nu_e$ oscillations and L (Kirilova, 2018). It was found that

$$L > (0.01\delta m^2 / eV^2)^{3/5}$$

may suppress $\nu_s \leftrightarrow \nu_e$ and eliminate BBN bounds on neutrino $\nu_s \leftrightarrow \nu_e$ oscillations The DR problem may be solved applying this mechanism if L is big enough, namely : L>0.074.

5. BBN CONSTRAINTS ON BSM WITH CHIRAL TENSOR PARTICLES

Chiral tensor particles (CTP) have been proposed on theoretical grounds as an extension of Standard Model for completeness of the representation of the Lorentz group by Chizhov (1993). For more detail about the extended model with

CTP see also Chizhov (2011). Experimental search for these particles is conducted at present at the ATLAS experiment of the Large Hadron Collider, where first experimental constraints on their characteristics were obtained (Aad et al., 2014, 2014a).

The cosmological place of CTP was studied in several publications, see for example Kirilova et al. (1995), M. Chizhov&Kirilova (2009); Kirilova &V. Chizhov(2017), Kirilova&E.Chizhov(2018).Their characteristic interactions in the early Universe plasma, namely, their creation, scattering, annihilation and decay processes were studied and the corresponding time and energy intervals of their effective presence were obtained. The period of CTP effectiveness was determined to be:

$$6.10^{-42}\ \text{s} < t < 6.5.10^{-27}\ \text{s}$$

CTP dynamical cosmological effect was studied also. Due to their contribution to the matter tensor in the right-hand side of the Einstein--Hilbert equation, CTP increase Universe density and change the dynamical evolution of the Universe. Namely, the energy density increase caused by the additional degrees of freedom g_{CTP}=28 in the BSM physics model with CTP leads to speed up of the expansion of the Universe during the period of CTP presence:

$$H = \sqrt{\frac{8\pi^3 G_N g_*^{new}(T)}{90}}\, T^2$$

where $g^{new}=g_{SM}+ g_{CTP}$.

The provided analysis of the cosmological place of the CTP showed, that cosmology allows their presence. Their direct interactions with the components of the high temperature plasma were effective for a very short early period during the Universe.

Recently reconsideration of the BBN constraint on the CTP coupling constant was provided (Kirilova&E.Chizhov, 2018). Using different BBN constraints on the additional relativistic particles and assuming that CTP interact with light sterile neutrinos, we have calculated the decoupling of right-handed neutrino production and obtained BBN cosmological constraints on CTP interaction strength for different cases. The constraints on CTP coupling in case of 3, 2 and 1 light ν_e and BBN constraint (Pitrou et al. 2018) $\delta N_{eff} < 0.3$, read correspondingly:

$$G_T \le 4.2\times10^{-4} G_F \qquad G_T \le 8.4.\times10^{-4} G_F \qquad G_T \le 1.4.\times10^{-3} G_F$$

Thus, BBN constrains CTP interaction strength to be milli weak or weaker, depending on the number of light right-handed neutrino species. These constraints can be interpreted also as an indication for absence of CTP interactions with light sterile neutrinos, or for absence of light sterile neutrinos.

CTP were present at energies typical for inflation, Universe reheating and leptogenesis and baryogenesis. Hence, it is interesting to explore further the role of CTP in the very early Universe.

6. CONCLUSIONS

Fruitful interplay between cosmology and particle physics exists. Cosmology can predict the influence of BSM characteristics and test them. In particular, BBN is the earliest and the most reliable and precision probe of BSM physics, relevant at energies typical for BBN (MeV-KeV). It measures neutrino mass differences, number of neutrino species, neutrino mixing patameters, deviations from equilibrium, baryon density and lepton asymmetry of the Universe, new interactions, additional particles, etc.

BBN is a reliable baryometer. The baryon density is measured with great precision and points to BSM physics – necessity of nonbaryonic dark matter. Its nature is still an open issue both in cosmology and in particle physics. Though baryon density is measured with a high accuracy today, the exact baryogenesis mechanism is not known. The problem of baryon asymmetry of the Universe is still fascinating and its explanation probably is in the realm of BSM physics. The possibility for astronomically large antimatter objects is experimentally and theoretically studied. This issue has gained more interest recently with the detection of anti He-3, -4 nuclei in cosmic rays. The separation mechanisms between matter and antimatter domains also imply BSM physics.

BBN is a very sensitive leptometer. Degenerate BBN constrains L : $|L|<0.1$, while in case of active-sterile neutrino oscillations L as small as 10^{-8} may be felt by BBN.

BBN is the most sensitive speedometer. It constrains additional light particle species N_{eff} thus, constraining SUSY, string, extradimensional models, etc. BBN bounds on N_{eff} are strengthened in case of neutrino oscillations.

BBN severely constrains CTP interaction strength in case CTP interact with light sterile neutrinos.

BBN constrains neutrino oscillations parameters. It provides the most stringent constraint on the neutrino mass differences δm^2. BBN constraints on neutrino oscillations parameters depend nontrivially on the population of sterile neutrino and L in the Universe. Additional initial population of the sterile state not always leads to strengthening of constraints (as can be naively thought) it may relax them. Relic L may provide relaxation or enhancement of BBN constraints on oscillations.

Large enough L may provide relaxation of BBN constraints on oscillations, by suppressing oscillations and causing incomplete thermalization of the sterile neutrino. Thus, dark radiation (1+3 oscillations models) might be allowed by BBN with L.

Future cosmic missions and observations and experiments at accelerators and colliders are expected to improve our knowledge about the Universe and, in particular, to solve the riddles of dark matter, dark energy, inflation, baryon asymmetry, lepton asymmetry, additional interactions and/or particles, dark radiation, etc. and hopefully detect the Nature chosen BSM physics.

Acknowledgements

I would like to thank the organizers for the kind invitation to present this invited talk at the XI Bulgarian-Serbian Astronomical Conference. The work on this publication was partially supported by projects DN08-1/2016 and DN18/13-12.12.2017 of the Bulgarian National Science Fund of the Bulgarian Ministry of Education and Science.

References

Aad, G. et al.(ATLAS Collab.): 2014, *Phys.Rev. D* **90**, 05, 205.
Aad, G. et et al. (ATLAS Collab.): 2014a, *JHEP* **09**, 037.
Ade, P. et al. (Planck Collaboration): 2016, *Astron. Astrphys.* **594**, A13.
Angulo, C. et al.: 1999, *Nucl. Phys. A* **656**, 3.
Arbey, A.: 2012, *Comput. Phys. Commun.* **183**, 1822.
Aver, E., Olive, K., Skillman, E.: 2015, *JCAP* **7,** 011.
Barbieri, R., Dolgov, A.: 1990, *Phys. Lett. B* **237**, 440.
Barbieri, R., Dolgov, A.: 1991, *Nucl. Phys. B* **349**, 743.
Chizhov, M. V.: 1993. *Mod. Phys. Lett. A,* **8**, 2753.
Chizhov, M. V.: 2011, *Sov. J. Part. Nucl.,***42**, 93.
Chizhov, M. V., Kirilova, D.: 2009, *Int. J. Mod. Phys. A* **24**, 1643.
Coc, A. et al.: 2007, *Phys. Rev. D* **76**, 023511.
Coc, A. et al.: 2012, *Phys. Rev. D* **86**, 043529.
Consiglio, et al.: 2017, *arXiv* 1712.04378.
Cooke, R., Pettini, M., Steidel, C: 2017, *arXiv* 1710.11129.
Cyburt, R., Fields, B., Olive, K., Yeh T.-H.: 2016, *Rev. Mod.Phys.* **88**, 015004.
Capozzi, et al.: 2017, *PRD* **95**,033006.
Danforth, C., Shull, M.: 2008, *Astrop. J.,* **679**, 1, 194.
Dmitriev, V., Flambaum, V., Webb, J.: 2004, *Phys. Rev. D* **69**, 063506.
Dolgov, A.: 1981, *Sov. J. Nucl. Phys.* **33**, 700.
Dolgov, A.: 2002, *Phys. Rept.* **370**, 333.
Dolgov, A., Kawasaki, M., Kevlishvili, N.: 2009, *Nucl.Phys.B* **807**, 229.
Dolgov, A. et al., 2002, *Nucl. Phys B* **632**, 363.
Enqvist K., Kainulainen K., Thompson M.: 1992, *Nucl. Phys. B* **373**, 498.
Izotov, Y., Thuan, T., Guseva, N.: 2014, *MNRAS* **445**, 778.
Foot, R., Volkas, R.: 1995, *Phys. Rev. Lett.* **75**, 4350.
Foot, R., Thomson, M., Volkas, R.: 1996, *Phys. Rev. D,* **53**
Gariazzo, S., Giunti, C., Laveder, M., Li, Y. F.: 2018, *arXiv:*1801.06467v3.
Giunti, C.: 2017, *Nuclear Particle Physics Proceedings* **287**, 133.
Hannestad, S., Tamborra, I., Tram, T.: 2012, *JCAP* **1207**, 025.
Kawano, L.: 1992, *Let's go: Early Universe2 Primordial Nucleosynthesis: The Computer Way.*
Kirilova, D.: 2018 (to be published)

Kirilova, D.: 2013, *Hyperfine Interact.* **215**,111.
Kirilova, D.: 2012, *JCAP* **06**, 007.
Kirilova, D.: 2007, *IJMPD* **16**, 7, 1.
Kirilova, D.: 2004, *Int. J. Mod. Phys. D* **13**, 831.
Kirilova, D.: 1988, *JINR* preprint E2-88-301.
Kirilova, D., Chizhov, E.: 2018 (to be published)
Kirilova, D., Chizhov, V.: 2017, *IJ Modern Physics Letters A* **32**, 1750187.
Kirilova, D., Chizhov, M.: 2009, *Proc. XXIth Rencontres de Blois "Windows on the Universe"* eds L. Celnikier, J. Dumarchez, and J. Tran Thanh Van, Moriond Astrophysics Meeting, The Gioi Publishers, Vietnam GPXB 4 - 1000/XB-QLXB.
Kirilova, D., Chizhov, M.: 1996, in *Neutrino* **96**, 478.
Kirilova, D., Chizhov, M.: 1997, *Phys. Lett. B* **393**, 375.
Kirilova, D., Chizhov, M.: 1998a, *Phys. Rev. D* **58**, 073004.
Kirilova, D., Chizhov, M.: 1998b, *Nucl. Phys. B* **534**, 447.
Kirilova, D., Chizhov, M.: 2000a, *MNRAS* **314**, 256.
Kirilova, D., Chizhov,M.: 2000b, *Nucl. Phys. B* **591**,457.
Kirilova, D., Panayotova, M.: 2006, *JCAP* **12**, 014.
Kirilova, D. P., Chizhov, M. V., Velchev, T. V.: 1995, *Comptes Rendus de l'Académie bulgare des Sciences* **48**, 25.
Lesgourgue, J., Pastor, S.: 1999, *Phys. Rev. D* **60**, 103521.
Mangano, G., Serpico, P.: 2011, *PLB* **701**, 296.
Mangano, G., Miele, G., Pastor, S., Pisanti, O., Sarikas, S.: 2011, *JCAP* **11**, 035.
Mirizzi, A., et al.: 2012, *PRD* **86**, 053009.
Partignani, C., Particle Data Group: 2016 and 2017 update, *Chineese physics C* **40,** 100001.
Pisanti, O. et al.: 2008, *Comput. Phys. Commun.* **178,** 956.
Pitrou, C., Coc, A., Uzan, J.-P., Vangioni, E.: 2018, *arXiv*:1801.08023
Rossi, G., Yeche, C., Palanque-Delabrouille, N., Lesgougues, J.: 2015, *PRD* **92**, 063505.
Sasankan, N., Gandopadhyay, M. R., Mathews, G. J., Kusakabe, M.: 2017, *Int. J. Mod. Phys. E* **26**, 7 , 1741007.
Sbordone, L. et al.: 2010, *Astron. Astrophys.* **522**, A26.
Serebrov, A. et al.: 2017, arXiv 1712.05663.
Shvartsman, V.: 1969, *JETP Lett.* **9**, 184.
Steigman, G.: 2012, *Advances in High Energy Physics,* **2012**, 268321.
Wagoner, R., Fowler, W., Hoyle, F.: 1967, *Astrophys. J. Supp.* **148**, 3.
Xu, Y. al.: 2013, *Nucl. Phys. A* **918**, 61.

Contributed Papers

Proceedings of the XI Bulgarian-Serbian Astronomical Conference (XI BSAC) Belogradchik, Bulgaria, May 14-18, 2018
Editors: Milcho K. Tsvetkov, Milan S. Dimitrijević and Momchil Dechev
Publ. Astron. Soc. "Rudjer Bošković" No 18, 2018, 39-52

FELIX ROMULIANA – AN ARCHAEOASTRONOMICAL RESEARCH

ALEKSANDRA BAJIĆ[1] and MILAN S. DIMITRIJEVIĆ[2]

[1]*Society for archaeoastronomy and ethnoastronomical research, "Vlasici", Belgrade, Serbia*
[2]*Astronomical Observatory Belgrade, Serbia*
E-mail: aleksandra.bajic@gmail.com, mdimitrijevic@aob.rs

Abstract: This paper is an attempt to determine the festivals celebrated in Felix Romuliana using archaeo-astronomical methods. Some deities considered protectors of its inhabitants are known. Some iconographic artworks were found at the site in the form of sculptures, reliefs or mosaics, pointing to the worship of Dionysus and Ariadne, Jupiter, Hercules and Aesculapius and probably Demeter and Hecate. The assumption that their festivals were determined by observing the Sun and the stars was checked (after a geodetic survey of the horizon) using the astronomical software RedShift 7 and PLSV. The observation points were determined according to the rules of Roman architecture. The mutual relations of the buildings of the sacred funeral complex were examined and a number of astronomically important directions were found.

1. INTRODUCTION

Felix Romuliana is a late-ancient complex of buildings, built during the era of tetrarchy. It was constructed by the order of Gaius Galerius Maximianus Augustus, the Roman Emperor (*c*. 260 – 311 AD), at the placc of his birth, near today's Gamzigrad, in the eastern part of Serbia, and named after his mother, Romula. The origin of this Roman emperor is not Roman: a father whose name is not preserved was a Thracian while his mother was a Romanized Dacian.

The complex of buildings in Felix Romuliana consists of two parts: fortification with two residential palaces, two temples, together with a bath (Termae) and a granary; the other part is a sacral and funeral complex, located east of the fortification, on the hillock named Magura, with two tumuli, two mausoleums, an insufficiently studied and unpublished building and a Tetrapilon. Archaeological research was carried out by a team of archaeologists led by D. Srejović.

Roman planning of cities and military camps followed some strictly defined rules[1]. According to Vitruvius, the north-south direction (lat. Cardo) was to be determined first. After that, east-west direction was determined, named Decumanus. Two main streets of the city (or the camp) followed these directions, named Cardo Maximus and Decumanus Maximus. The main geodetic point of the city, called *GROMA* or *Umbilicus* ("navel") was at the juncture of these two streets. As a rule, the main gate of the city was in the east, called *Porta Praetoria*. There was also a second gate, in the west, called *Porta Decumana*. The latter most often led to the sacral-funeral complex, which was located west of the city (or the camp). Depending on the configuration of the terrain, it may have been possible to deviate from these rules to a certain degree, but never significantly.

At first glance, it is obvious that the plan of Felix Romuliana deviates significantly from these rules: first, the north-south direction (*Cardo*) can not be identified in the residential complex. There is only one street, *Decumanus Maximus*, following east-west direction. There is no intersection of the two main streets. So, the position of *GROMA* is quite uncertain. This main geodetic point was usually marked by a flat, square stone, with a carved line on its upper surface pointing to the north, and such a stone was not identified at the site. The second deviation refers to the demolition of the older fortification followed by the construction of a new one. Knowing that fortifications were traditionally considered sacred by the Romans, it is quite unclear what reasons could have led Galerius to do so. Why was the place of the western gate of the complex changed? It is also unusual that the sacral-funeral complex is to the east of the residential one. The existence of tumuli at the Roman site is also unique. Tumuli were usually built by the Thracian, the Dacian and the Macedonian rulers, and not by Roman emperors. Should the reasons for these deviations be sought in the non-Roman emperor's origin? Are some non-Roman traditions together with some local, non-Roman knowledge embedded in the foundations of Felix Romuliana?

2. WHERE IS THAT GROMA (UMBILICUS)?

As Cardo is absent, Groma could be somewhere on the *Decumanus* road. The eastern gate is located on a low position, it is 11 m lower in respect to the western one. It is not practical for Groma to be there. On the contrary, the original Western Gate (*Porta Decumana*) is the highest point of the complex, the view to the terrain is excellent as seen from there. The place is quite convenient to be the main geodetic point. This hypothesis was the first one tested in this paper by the archaeoastronomical method, in order to discover some sense of its position. The Google Earth and Zoom Earth satellite maps were used to identify geographical coordinates of points of interest. To transform the coordinates into all three systems, the Geotrans and Coordinate transformer Chuck Taylor software were

[1] Vitruvije, *Deset knjiga o arhitekturi* (2014, prevod Matija Lopac 1951), Orion art, Beograd (Vitruvius, *Ten books on architecture*).

used. Using of WGS 84/UTM coordinate system, which expresses the geographic coordinates in meters, gives the possibility of calculating the azimuths of directions and the distances between the individual points; data on altitudes are available on the maps of the site created by Dragoslav Srejović's team, as well as on the maps of Military Geographical Institute of Serbia. These data allow the calculation of altitude differences and angular heights of certain points on the horizon. Astronomical software Red Shift 7 and PLSV (Planetary, Stellar and Lunar Visibility) were used to define astronomical events in the given directions. In the end, a geodetic survey was performed from the hypothetical *Groma*, in order to check the calculated values of azimuths and the angular heights of some distinctive points on the horizon.

Figure 1: Groma and the directions towards the structures on the natural horizon.

The hypothetical *Groma* is located on the following coordinates: **43° 53' 57.33''N** 22° 11' 5.41'' E. It was noticed the northernmost point of the smaller tumulus is at the same latitude (**43° 53' 57.33''N** 22° 11' 59.92'' E). These two points define the *Decumanus* of the complex. The eastern horizon on Magura is well structured: the trough which contains the remains of Tetrapilon is visible; both *tumuli* and Mausoleum 1 (the Mausoleum of Romula, the emperor's mother) are clearly visible; Mausoleum 2 (the Mausoleum of Galerius) is obscured by Tumulus 2. About 200 m north of the smaller tumulus, there are the remnants of a big rectangular building of an unexplained purpose, which for now is known to have been richly decorated with mosaics. These remnants are not visible being clogged with earth for the sake of protection, but their position is known - archaeologists have recorded the exact coordinates of the building's corners in the WGS 84 / UTM system, so the azimut of its direction can be calculated.

Geodetic survey was performed from the hypothetical Groma in November 2017. and the following results were obtained:

	azimuth	altitude
Tetrapilon (south)	**79° 08' 40''**	**01° 43' 30''**
Unpublished building(south)	84° 21' 36''	02° 41' 19''
Tumulus 1 (north)	**90° 00' 00''**	**02° 51' 10''**
Tumulus 1 (south)	**91° 09' 20''**	**02° 50' 40''**
Between the Tumuli	**91° 17' 40''**	**02° 50' 30''**
Tumulus 2 (north)	**91° 31' 50''**	**02° 50' 00''**
Tumulus 2 (south)	**92° 54' 30''**	**02° 44' 50''**

* The calculated values are marked gray, the measured values are marked black.

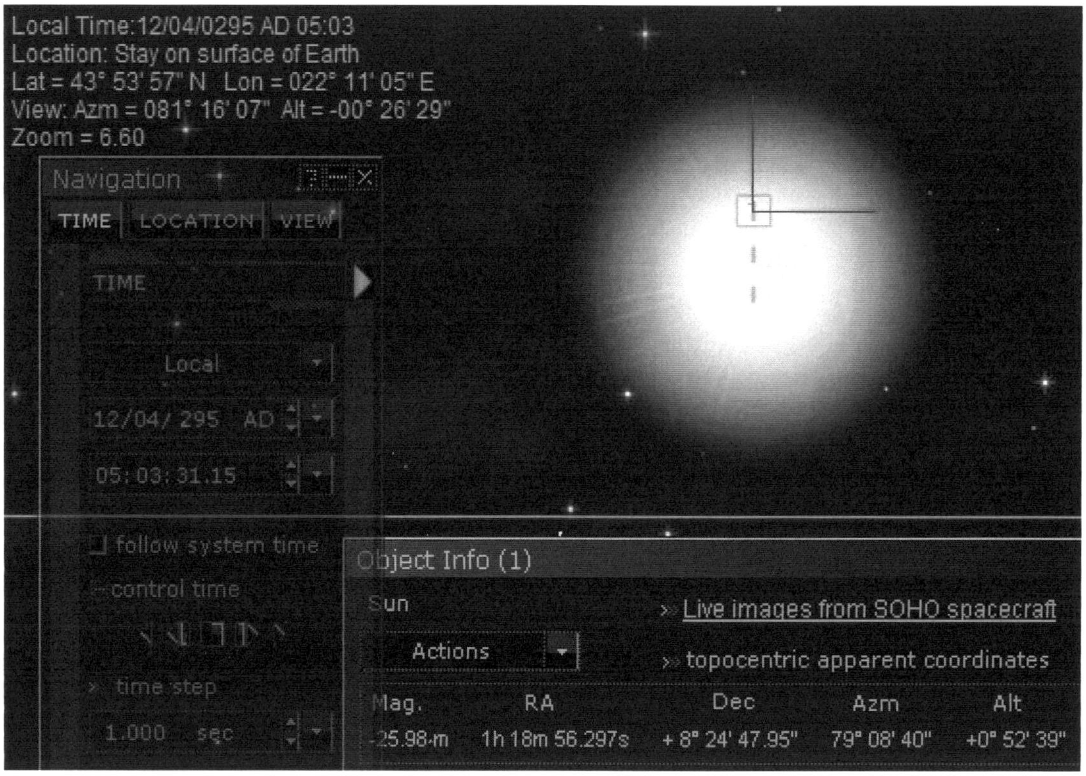

Figure 2: The Sun rise at the beginning of Cerialia on RedShift 7.

Atmospheric refraction has been taken into account. Its average values were used for each month of the year, according to the measurements of Sampson and associates (2003)[2]. The software takes the proper atmospheric refraction on the horizon automatically only if it is plane.

As seen from Groma, the first sunrays appeared at the direction of the right (southern) foot of **Tetrapylon** on April 12, denoting ***Cerialia***, the holidays

[2] Sampson, R. D; Lozowski, E. P; Peterson, A. E; Hube, D.P; 2003, Variability of the Astronomical Refraction of the Rising and Setting Sun, *Publications of the Astronomical society of the Pacific* **115**, 1256 – 1261.

dedicated to Ceres, the goddess of crops. A fragment of the statue depicting Ceres (a hand holding a torch), was identified at the site. According to the famous myth, the goddess held a torch while she was searching her Proserpine, accompanied with Hecate, the protector of crossroads. Tetrapylon is located on a crossroads.

The first sunrays appeared at the southern point of Tumulus 2 on March 19, marking the beginning of ***Quinquatrus***, the holiday of Minerva, the goddess of war and strategy. Galerius the Emperor was both: a warrior and a strategist. Minerva was to be worshiped.

When the Sun rose at the northern point of Tumulus 2, *Qinquatrus* ended and *Tubilustria* begun.

As seen from *Groma*, the Sun rose at the direction of the unpublished building on April 4, the day of Megalesian Cybele, the protectress of fortifications.

In 307.AD, Galeria Valeria was honoured with the title *„Mater Castrorum“,* ("Mother of fortifications") suggesting her identification with Cybele, the goddess. One assumption could be that the unpublished building was intended to be the Mausoleum of Valeria Galeria, the emperors wife.

According to the Orphic myth, when the first Dionis, Zagreus (the son of Zeus and Persephone) was born, Zeus was aware of the danger that threatens the divine boy from his jealous stepmother, Hera. So he gave him to Hermes, who took the child to Mount Ida and handed him over to Cybele[3]. Thus, the goddess became the protector of Dionysus (Zagreus).

This position of the Sun, that marks the Cybele's festival, was determined by its centre, unlike the previous two cases, where it was determined by its first rays. This could have some significance, and will be discussed later.

Thus, the hypothetical *Groma* gets its astronomical meaning, as the point from which sunrise was observed in order to determine certain Roman festivals. Ovid reported in detail about the position of festivals in the Roman calendar in his work *Fasti*, giving his readers clear explanations that many of those were determined according to astronomical events.

Felix Romuliana is far from Rome. There were no libraries available there either, and the announcements of the high Roman clergy couldn't quickly be transferred to the place. Its inhabitants apparently had to determine their own festivals by themselves.

It is interesting that all the festivals determined by observing the Sun from the *Groma*, are dedicated to female deities. Likewise, it is notable that not all surveyed points on the horizon (covered by geodetic survey) were relevant, but only three of those. It could therefore be assumed that some of these points are important for some other observation points located in the residential complex. Temples are the first free association. Therefore, the following question was raised:

---

[3] Nonnus, *Dionysiaca* 9. 136 ff :

3. WHAT WAS TO BE OBSERVED FROM THE TEMPLES OF FELIX ROMULIANA?

3.1. The bigger temple

Within the fortification with the residential complex, two temples were identified: the larger one is south of *Decumanus Maximus*. It was considered to be dedicated to Jupiter, the supreme Roman deity. The smaller one is located in the northern part of the fortification, close to its northern wall, presumably being dedicated to Liber and Libera (the Roman equivalent of Greek Dionysus and his wife Ariadne). The first assumption is based on historical facts: Galerius was the adopted son of Diocletianus who was considered the earthly incarnation of Jupiter.

Figure 3.

It was Diocletianus who founded the system of tetrarchs in Roman Empire, adopted Galerius as his son and appointed him as Caesar in 293 AD. After his withdrawal from power in 305 AD, he gave his adopted son the title of Augustus. Jupiter's cult is attested in Felix Romuliana: the marble head of Jupiter and his left foot were found at the site. Based on the size of these remnants, the statue was about 6 m high.

It is not allowed and it is not possible to climb into the larger temple and to install a theodolite there. So the examination of the temple was done by identifying the central point at the back of the hall, to the opposite of the doors. Its coordinates are; 43° 53 '56.26"N 22° 11' 09.88 " E. The axis of the temple is on the azimuth A = 86° 56'. Looking in that direction one can see the northern foot of the Mausoleum 1 (the Mausoleum of Romula). The calculated height of the horizon is h = 3° 24".

According to Red Shift 7, the Sun rose on the given azimuth on September 13th, the day of Ides and the day of ***Epulum Iovis*** – The feast of Jupiter. Its upper left quarter was observed on the left (northern) foot of Romula's Mausoleum:

Symbolically, Jupiter, taking the form of the Sun (the golden dragon) „visited" Romula on the day of his feast. Just like he „visited" Proserpine (Persephone) in the Orphic myth, in the form of a dragon, conceiving Dionysus Zagreus, the first incarnation of Dionysus (Liber).

The position of the Sun was determined by its centre again. This could have some significance, and will be discussed later.

According to Lactantius, Galerius was spreading the story that his mother had been fertilized by a dragon to give the emperor's birth. Such story would be streghtened if the day of Galerius' birth, which is not known, was nine months after the Ides of September, i.e. about the Ides of Jun.

Whatever, This story of Lactantius may be first indication that the emperor was identified with Liber (Dionysus).

The cult of Liber (Dionysus) was richly attested archaeologically in Felix Romuliana by the discovery of a magnificent mosaic, with the image of the deity. His divine wife Libera (Greek Ariadne) is also present on the relief depicting her figure. It's the same Ariadne, the daughter of Minos from Crete, who gave Theseus the famous thread to get out of the labyrinth in which he killed Minotaur, the monster with a bull's head. After that event, they fled from Crete together, as Theseus promised to marry her. They stopped on the island of Naxos to restore their water supplies. According to the myth, Ariadne had fallen asleep and Theseus fled away towards Mycenae leaving her on the island. There she was noticed by Dionysus. The god fell in love, and she became his wife. Of course, a mosaic with a maze was also discovered in Felix Romuliana.

3.2. The smaller temple

It is possible to enter the smaller temple, its staircase is preserved. But it is not allowed, so the examination was done in a similar way as with the bigger one. The coordinates of the central point at the depth of the temple (to the opposite of its entrance) are given in the picture. The axis of the temple is on the azimuth A = 94° 36' and the calculated height of the horizon in that direction is h = 3° 22'.

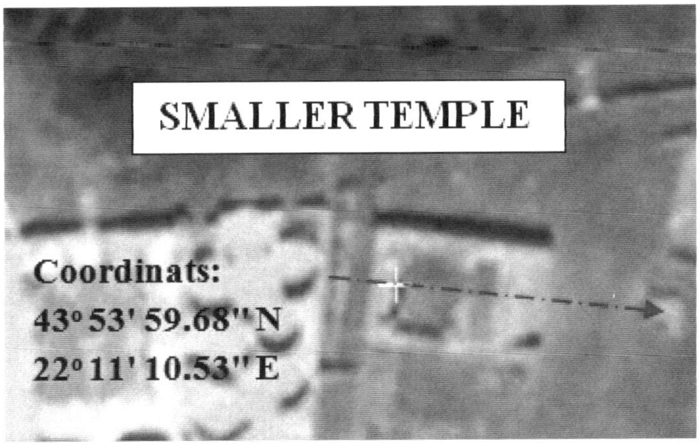

Figure 4.

On March 17, the day of ***Liberalia***, the upper pole of the Sun rose exactly between the two tumuli, above the altar, as seen from the smaller temple.

The two tumuli are the most impressive structures on the eastern horizon. Such a sunrise had to look magnificent. On March 17th 2018. we went to the site to take a photo of it, but we were not lucky: the sky was cloudy. Anyway, the assumption that the smaller temple was dedicated to Liber (Dionysus) becomes more firmly founded.

All these results lead to the following question: Was the sacred-funeral complex on Magura[4] also measured by celestial measures? Today, this hill is called Zvezdan, just like the nearby village. The name is derived from the Slavic (Serbian) adjective, which means "starry". The fact that Magura was the place for star gazing is still preserved in the local toponomastics.

3.3. The sacral and funeral complex

What was once called Magura is a flattened hillock, 252 m high. There are two *Tumuli* on its top, together with two Mausoleums. Other buildings on the same hill are on its slopes, at much lower elevation. *Tumuli* and Mausoleums are at the most dominant features of the hill. The smaller, northern Tumulus is 36 m in diameter. It is 4-5 m high, enclosed by a stone fence about 1.5 m preserved height. Mausoleum 1 is next to it, presumed to belong to Romula. The larger, southern Tumulus is 40 m in diameter, about 1-2 m higher in respect to the first one. Mausoleum 2 is in the vicinity, presumed to belong to Galerius. Both the Mausoleums are poorly preserved, the foundations and core of the base only. The base of Mausoleum 1 was in the shape of a four-pronged prism with an octagonal *Coela* above. The base of Mausoleum 2 was a prism with twelve angles, crowned by 12 pillars surrounding a cylindrical *Coela*. According to the proposal of ideal reconstruction, the whole complex looked like in Fig. 9.

Both tumuli had been the places of funeral pyre[5] and were fully archaeologically explored. A Bronze Age necropolis[6] was discovered below the Roman layer, which means that the site had its sacral importance even then.

Before the ritual of the apotheosis (cremation) of the imperator's mother was performed, the northern tumulus had not been clogged with earth, and it certainly

---

[4] According to Romanian researcher Jon I. Russu / Ion I. Russu, (1967) Limba traco-dacilor, Editura Ştiinţifica, Bucuresti /, Magura is a word from the Dacian language, meaning "hill" or "hillock". There is a cave with the same name in Bulgaria, which contains prehistoric drawings, considered to be "notes" of celestial phenomena. There is a village with the same name in today's Romania, near the town named Bran (and Bran is the name of the Celtic solar deity), as well as the village of Magura in Kosovo, west of Lipljan, near the Roman Ulpiana. Thus, the term "Magura Hill" is actually a pleonasm.

[5] Popović, Ivana i grupa autora, (2010) *Felix Romuliana – Gamzigrad,* Arheološki institut, Beograd, Posebna izdanja 47. (the article: Ivana Popović, *Sakralno-funerarni kompleks na Maguri*, p. 141)

[6] Idem.

existed because its northern and southern points were the important landmarks for the construction of the residential complex structures. The northern one defined the *Decumanus*, so it had to be established very early, practically before the construction of any other building. Therefore, given the circular shape of its fence, it can be assumed that it was originally the place for sky observing in the conditions of a flat arteficial horizon, provided by the stone fence reaching the height of the observer's eyes.

Figure 5: Ideal reconstruction of the sacral-funeral complex on Magura, a view from the north.

A similar assumption can be applied to the second, southern tumulus, which could have the same function. This hypothesis can be examined in a similar way as it was done with the temples in a residential complex. It can be imagined that there are no tumuli but two circular formations with fences only, together with two buildings intended to be Mausoleums (or at least the foundations of these).

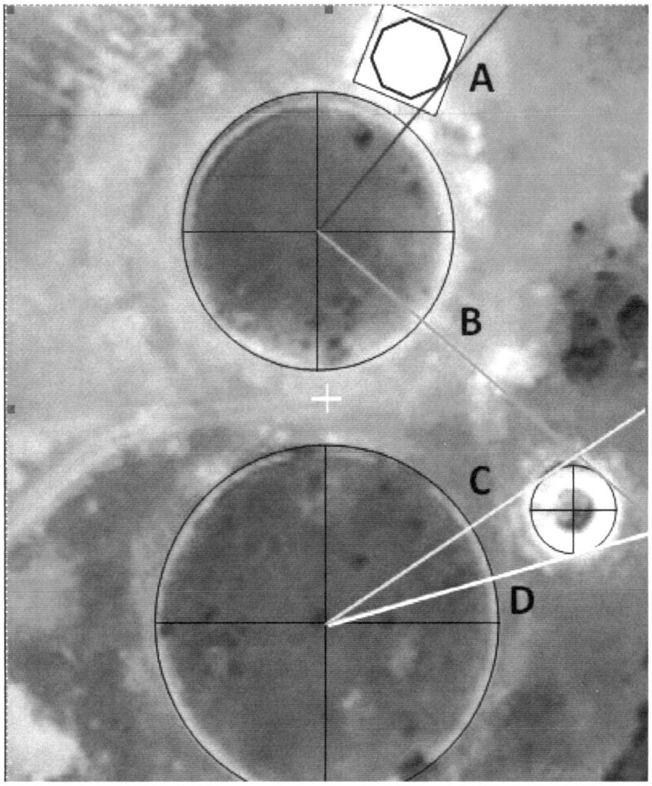

Figure 6: The picture shows four well established directions (A,B,C,D) that define the relationship between circular formations and Mausoleums. They are marked with different colours.

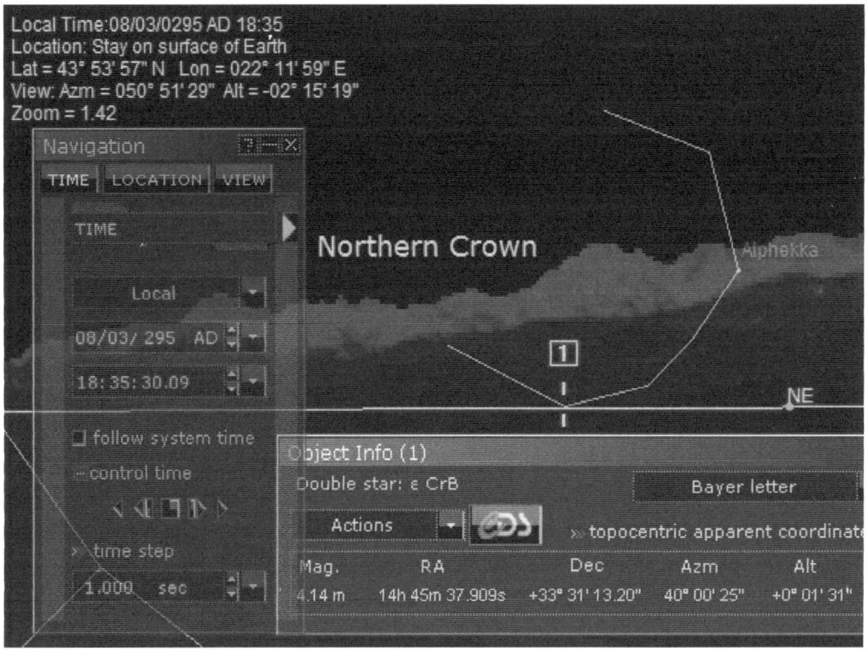

Figure 7: This is the rising direction of Corona Borealis, the constellation known since the ancient times. This constellation had its significance in the cult of Liber and Libera (Ariadne, the Dionysus' wife).

According to the myth, it was the crown, given to Ariadne as the wedding gift, when she married Dionysus. It was forged by Vulcan (Greek Haephaestus) and was ascended to the heaven as the memory of the happy event.

According to the astronomical software PLSV (Planetary, Stellar and Lunar visibility), In the third century AD, Corona Borealis[7] had its evening rise exactly on March 17th, the day of ***Liberalia***, the main holiday of Liber and Libera.

The direction B on the picture is a tangent from the centre of the northern tumulus to the basis of the Mausoleum 2. It is on the azimuth A = c. 131° and the horizon altitude is h = 0°

It was a quite unexpected finding: Southern Major Lunar Standstill moonrise. The Moon's first rays appeared by the left (north-eastern) foot of the Mausoleum 2, as seen from the centre of the smaller circle. Then it was partly obscured by the Coela while some of its surface was visible between the pilars and Coela. After that, it rose again on the other (south-western) side of the Mausoleum 2.

There is no evidence that Lunar standstill had any significance in the Roman religion. It was important for Bronze age Celts. But, Celts (Scordisci) lived in this area before the Roman conquest. They mixed with the indigenous Thracians and Dacians, forming the tribe named Pikensians*, mentioned in historical records to inhabitate the area in the 2nd century AD. So, it is possible that the knowledge on Lunar Standstills thus became familiar to the local priesthood.

According to Red Shift 7, the astronomical software, 295. AD was the year of the Major Lunar Standstill which happened on June 14.

---

[7] The name of the Northern Crown in Dacian language is not known. However, there is one assumption that needs to be examined. The ancient Balts called this constellation Daržėlis (which the researchers translated as the "garden"). In the Hellenistic period of Greek history, in Odesos (now Varna, Bulgaria), the Thracians and Dacians worshiped the deity named Darzelas (or Darzalas), whose name was noted on the numerous coins minted in Odesos. The deity was considered chthonic, the bringer of abundance, fertility and masculinity. There was also a temple dedicated to him in Histria (today's Constanta in Romania). In his honor, games were held every four or five years (Darzeleia). According to Wikipedia (article: Thracian language), the word "zelas" means "wine", so it is possible to link Darzelas with Dionysus. The Old Balts' name for the asterism Hyades was Dievo Daržėlis, and the Hyades were the nymphs that nurtured Dionysus on the Mount Nisa. (*The Cosmology of the Ancient Balts*, authors: Straižys, V. & Klimka, (1997) L. *Journal for the History of Astronomy, Archaeoastronomy Supplement*, Vol **28**, pp. 57-80) The past paleolinguistic research assumes the connection of Dacian, Thracian and Baltic languages.

*Popović, Ivana i grupa autora, (2010) *Felix Romuliana – Gamzigrad,* Arheološki institut, Beograd, Posebna izdanja 47. (The article: Miodrag Sladić, *Gamzigrad u protoistoriji,* p. 31)

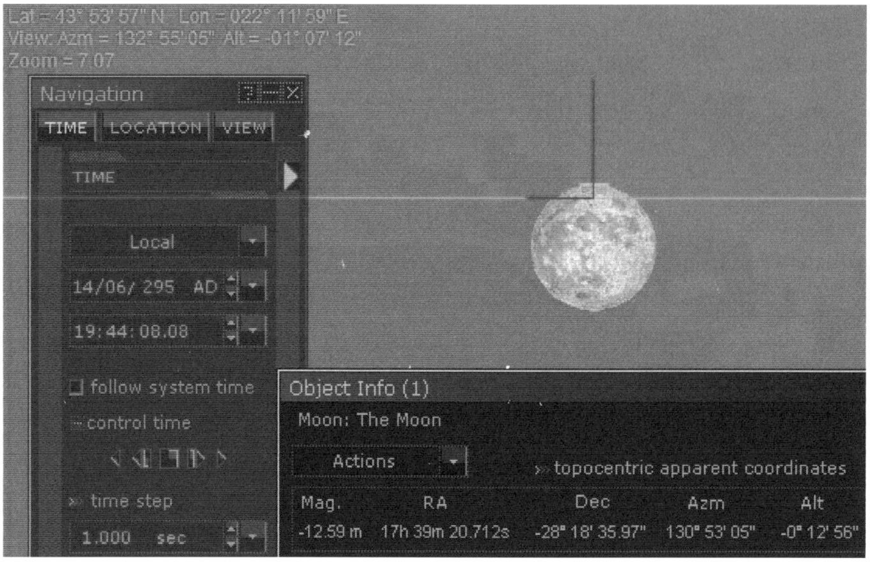

Figure 8: Lunar standstill.

Dragana Radovanović, the archaeologist, states that the probable year of Romula's death was 295AD. She found a collection of Roman gold coins at the place of Romula's cremation (smaller circular formation) and none of these coins were minted after 294. AD. Did someone find a way to make a record of Romula's death date astronomically? If it is so, then his record is very hard to extinguish, either by robbers or by the ravages of time.

But, there is another possibility, as the blue line makes a connection of the place of Romula's funeral pyre with Galerius' Mausoleum. That could also be the note on Galerius' birth. The same astronomical event happened in 258. AD, on June 3rd. As it is already mentioned, the emperor spread the story that his mother had been fertilized by a dragon, probably on the day of *Epulum Jovis* (the Ides of September) to give the birth of her son. So, the birth is supposed to come nine months later, about the Ides of June. The other directions on Magura have to be tested.

The direction C is on the azimuth $A = c. 55.5^{o}$ $h = 0^{o}$. It is the azimuth of summer solstice sunrise on June 21. 295, in terms of a flat horizon, as seen from the centre of the larger circular formation. The upper left quarter of the Sun appeared by the northern foot of the Mausoleum 2. It was not necessary to apply the correction for the atmospheric refraction on the horizon. It was applied automatically by the software itself, as the horizon is plane.

In his poem *Fasti,* (book VI), talking about this date, Ovid told the myth about the death and resurrection of Hypolitus. It is indeed a very convenient story for a funeral complex. Life and afterlife, death and apotheosis...

Now, it can be assumed that the direction of the southern lunar standstill was the last one determined from the centre of the smaller circular formation, on June 14. 295. AD. It had probably been done right after Romula's death, before the ritual of apotheosis was performed.

There is only one **direction D** left. It is on the azimuth $A = 73^o$-73.5^o and $h = 0^o$. It was the rising azimuth of Hyades, the asterism in Taurus in the trird century AD.

According to the myth, Hyades were sisters. They were the nurses of Liber (Dionysus) when he was a child, hidden on the Mount Nyssa, to be saved from Junona's (Hera, his stepmother) rage. So, the asterism could have its significance in the cult of Liber.

According to PLSV, the astronomical software, the apparent morning rise of Aldebaran, the brightest star in the asterism, was observed on June 11th, during the life span of Galerius. This date was the holiday named ***Matralia***.

It was dedicated to *Mater Matuta*[8], also known as Ino, Leucothea or Galatea, the sister of Semele, Dionysus' mother. She was also one of the nurses of Liber, his aunt and his stepmother. So, if Galeius was born in June 258. AD, he had an additional point to strenghten the believe of his connection with Liber.

We note as well that according to Lactantius[9] the Christian writer, Romula "organized sacrificial feasts with her peasants" very often. If she did it "with her peasants", she was often somewhere in a village. Felix Romuliana is imposed as that village, because it is the place where the peasants are "hers". This leads to the assumption that there was a local clergy, to which she may had been a member, taking into account her high hierarchical position of the emperor's mother. The feasts had to be held before 295, because it was the year of her death. At that time, there had to be a small village villa for her in Felix Romuliana, and two circular formations on Magura had been already built. If it is assumed that the construction works started in 293, it was a logical time to begin geodetic surveying and planning. As Galerius was very busy fighting and building his military and political career at the time, the only person of his confidence, who could have a direct control of the works, was his mother.

Really, this research has shown two types of astro-geodetic survey: the earlier one, empirical by its nature, relied on the first rays of the Sun. Considering that the author determined the direction of the major southern lunar standstill, it can be assumed that he was of local origin. It was the man of Romula's trust. The later one was probably done in 307. AD. It was oriented towards the centre of the solar disk, and its author probably also knew certain astronomical calculations. This second surveyor (and astronomer) determined three directions only: the direction of the sunrise on the Megalesian Cybele's day as seen from *Groma*, in order to determine the position of the Mausoleum for Valeria, the emperor's wife; the

---

[8] Ovidije, *Fasti*, (2016. Preveli i komentarisali Aleksandra Bajić i Milan Dimitrijević), Društvo "Vlašići", Beograd (Book 6).

[9] Popović, Ivana i grupa autora, (2010) *Felix Romuliana – Gamzigrad,* Arheološki institut, Beograd, Posebna izdanja 47. (the article: Aleksandar Popović, *Pisani izvori o Galeriju*, p. 215. The author explains that the word, translated as "paesants" was actually the Latin word "*vicani*", which means "villagers")

direction of the summer solstice sunrise, observed from the centre of a larger circle on Magura, in order to determine the definitive form of Galerius' Mausoleum; and the third, the direction of sunrise on the day of *Epulum Jovis*, in order to determine the exact position of the Temple of Jupiter.

Further research could help to find the exact order of construction of individual buildings of the complex, but this goes beyond the tasks of this paper.

References

Herodot, *Istorija* (1980, prevod Milana Arsenića), Matica Srpska, Novi Sad.

Hyginus, *Astronomica,* (prevod na Engleski jezik, Mary Grant), Theoi, Classical text library, http://www.theoi.com/Text/HyginusAstronomica.html [approached 10.10.2017].

Hyginus, *Fabullae,* (prevod na Engleski jezik Mary Grant) Theoi, Classical text library, http://www.theoi.com/Text/HyginusAstronomica.html [approached 17.10.2017].

Kelley, David; Milone, Eugene F; (2005) *Exploring Ancient Skies - An Encyclopedic Survey of Archaeoastronomy*; Springer-Verlag New York

Mladenović, Dragana: 2009, Astral path to Soul Salvation in Late Antiquity? The Orientation of Two Late Roman Imperial Mausolea from Eastern Serbia, *American Journal of Archaeology*, **113,** 81-97.

Ovidije, *Fasti*, (2016. Preveli i komentarisali Aleksandra Bajić i Milan Dimitrijević), Društvo "Vlašići", Beograd.

Popović, Ivana et al.: 2005, *Šarkamen (Eastern Serbia) - A Tetrarchic Imperial Palace and The Memorial Complex,* Archaeological Institute, Monography N° 45, Belgrade

Popović, Ivana et al.: 2010, *Felix Romuliana – Gamzigrad,* Arheološki institut, Beograd, Posebna izdanja 47.

Russu, Ion: 1967, *Limba traco-dacilor*, Editura Ştiinţifică, Bucureşti.

Srejović, D, Lalović, A., Janković, Dj: 1983, *Gamzigrad, kasnoantički carski dvorac (katalog izložbe)*, Galerija Srpske akademije nauka i umetnosti, Beograd.

Srejović, D., Vasić, Č.: 1994, *Imperial Mausolea and Consecration Memorials in Felix Romuliana, Gamzigrad, Eastern Serbia,* SANU, Beograd 1994.

Straižys, V, Klimka, L.: 1997, The Cosmology of the Ancient Balts, *Journal for the History of Astronomy, Archaeoastronomy Supplement*, **28**, 57-80.

Vasić, Čedomir: 1997, Simbolika sakralnog kompleksa u Feliks Romulijani, u *Uzdarje Dragoslavu Srejoviću: povodom šezdesetpet godina života od prijatelja, saradnika i učenika*, (editor Miroslav Lazić), Centar za arheološka istraživanja Filozofskog fakulteta, Beograd, 445-460.

Vitruvije, *Deset knjiga o arhitekturi* (2014, prevod Matija Lopac 1951.), Orion art, Beograd.

Proceedings of the XI Bulgarian-Serbian Astronomical Conference
(XI BSAC) Belogradchik, Bulgaria, May 14-18, 2018
Editors: Milcho K. Tsvetkov, Milan S. Dimitrijević and Momchil Dechev
Publ. Astron. Soc. "Rudjer Bošković" No 18, 2018, 53-58

DOPPLER IMAGING OF THE HERTZSPRUNG GAP STAR OU ANDROMEDAE

ANA BORISOVA[1], UWE WOLTER[2],
RENADA KONSTANTINOVA-ANTOVA[1] and K. P. SCHRÖDER[3]

[1]*Institute of Astronomy, Bulgarian Academy of Sciences,*
BG-1784, Sofia, Bulgaria
[2]*Hamburger Sternwarte, Gojenbergsweg 112, 21029 Hamburg, Germany*
[3]*Departamento Astronomia, Universidad de Guanajuato,*
GTO CP 36000, Mexico
E-mail: uwolter@hs.uni-hamburg.de

Abstract. OU And (HD 223460, HR 9024) is a moderately fast-rotating early G-type giant star, located in the Hertzsprung gap. As found by earlier Zeeman-Doppler imaging campaigns (ZDI), its surface magnetic field is dominated by a simple geometry: all Doppler images of OU And created so far show a polar spot, which appears to be stable at least on the timescale of several years. In combination with evolutionary models, this suggests that OU And has been an Ap-star during its main sequence phase. Our data cover close to ten years: 2008 to 2017. Using high-resolution spectra from two sites (NARVAL and TIGRE), we create photospheric Doppler images (DI) and, furthermore, analyze the evolution of OU And's chromospheric emission. Our analysis shows that, during all our observations, OU And has only shown feeble surface features apart from the pronounced polar spot. Furthermore, we find that in several instances these weak features move, evolve or disappear during less than one stellar rotation. We perform a systematic error analysis of the chromospheric activity indices on a subset of our TIGRE data. In this way, we find that our measurements of chromospheric activity vary weakly, but significantly on timescales of days up to weeks, but without any obvious rotational modulation. **In the framework of these proceedings we only outline our methods and results. For details, we refer to our article which has been submitted to the Bulgarian Astronomical Journal.**

1. INTRODUCTION

Our target OU And (HD 223460, HR 9024) is a single giant of spectral class G1 III (Gray et al. 2001) with moderate emission in the Ca II H&K lines (Cowley & Bidelman 1979, who did not further quantify the emission). It exhibits X-ray emission (Gondoin 2003, Ayres 2007) and is a moderate rotator with a projected rotational velocity of 21.5 km/s, making it an interesting target for Doppler imaging (DI) studies. Its position in the Hertzsprung-Russel diagram is in the Hertzsprung gap with an effective temperature of 5360 K and 2.85 solar masses.

The focus of our study is the characterization of OU And's activity on timescales up to several years. To this end, we use the longest time series of spectroscopic observations available for OU And to date - covering the years 2008 to 2017.

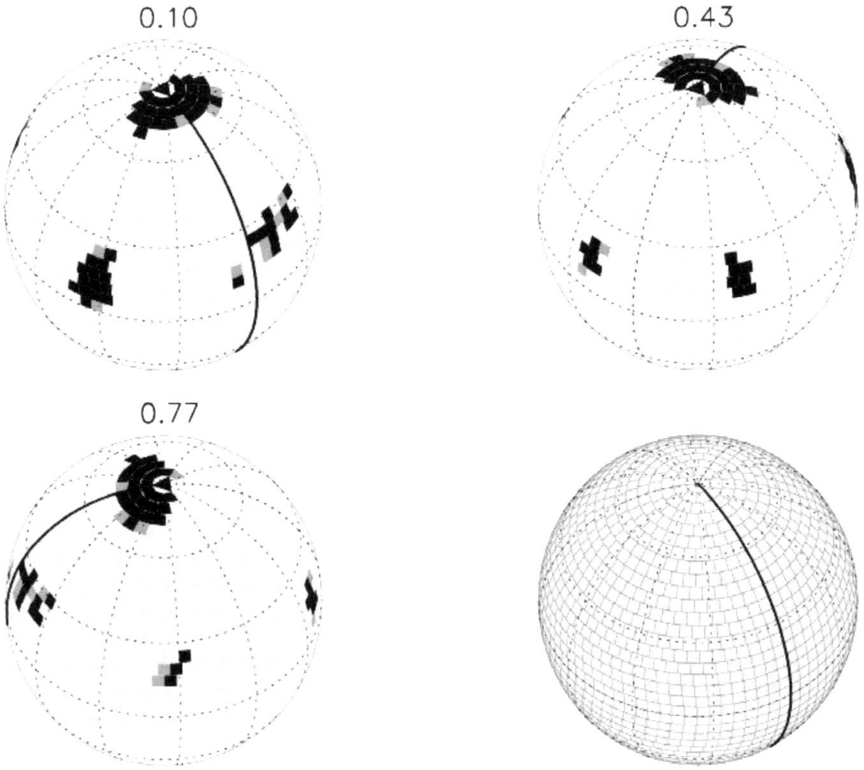

Figure 1: Photospheric Doppler image (DI) of OU And computed from our 2008 NARVAL spectra, using our DI package CLDI. All maps show the same surface for the indicated rotation phase; the sub-observer longitude at phase zero is marked by a thick line. The fourth map shows the empty surface grid used for our DI, it has 2048 elements, each extending by 4.5x4.5 square degrees. Black and gray elements indicate a spot filling factor of 100% and 50%, respectively.

2. OBSERVATIONS

Observations were carried out (i) with the NARVAL spectropolarimeter mounted on the 2 m telescope Bernard Lyot at Pic du Midi Observatory, France. Furthermore, (ii) we observed a long time series of spectra of OU And using the HEROS spectrograph of the 1.2m TIGRE robotic telescope at the La Luz observatory, near Guanajuato in central Mexico.

Using NARVAL, OU And was observed during the second semesters of 2008, 2013 and 2015. We observed 121 spectra of OU And using TIGRE/HEROS from November 2014 to September 2017, usually taking a few spectra per year. In addition, we performed one densely sampled campaign that covered nearly 3 stellar rotations during 14 observing nights from 2017 Sep 2 to 2017 Nov 14.

3. DOPPLER IMAGING THE NARVAL SPECTRA

To obtain spectral line profiles from the unpolarized (Stokes I) component of our NARVAL spectra, we use our least-squares deconvolution method, called "selective least-squares-deconvolution" (sLSD). Described in detail in Wolter et al. (2005), it uses comparatively narrow spectral regions, usually a few 10 Å wide to compute an average broadening profile of the lines in the region under study. We use our implementation of Doppler imaging, called Clean-like Doppler imaging (CLDI, see Wolter et al. 2005 & 2008) to reconstruct starspots on OU And's surface, i.e. cooler regions in its photosphere. Similar to the image deconvolution method CLEAN used in radio astronomy it performs an iterative reconstruction of the stellar surface, building up an increasingly good model of the observed line profiles.

The Doppler images of OU And were constructed with the same values of as those used for the ZDI imaging of Borisova et al. (2016, BEAL16). We made an attempt to independently estimate the rotation period and stellar inclination on the basis of the goodness of the line profile fit. Unfortunately, given the overall moderate rotation phase coverage of our spectra and - at times - fast spot evolution (compared to the rotation period) the CLDI tomography of our NARVAL data does not allow to determine the rotation period at all.

Based on the discussion of Doppler imaging caveats in our main journal article on OU And, we summarize here that we consider the polar spot, found in all our DIs of OU And, to be real. This means we are certain that it is not an imaging or line profile modeling artifact. See e.g. Bruls, Solanki & Schüssler (1998) for a detailed discussion of this aspect. Yet, the shape and size of the polar spot shown in our DI maps are only poorly defined, given the relatively small vsini, the potential uncertainties in stellar inclination, the partially uneven phase sampling and, maybe most importantly, the rather weak line profile deformations in our spectra of OU And. The same restrictions apply to the non-polar features of our DI: they are real in the sense of not being imaging or data artefacts. Yet, their precise positions, shapes and sizes are not well constrained by our DI.

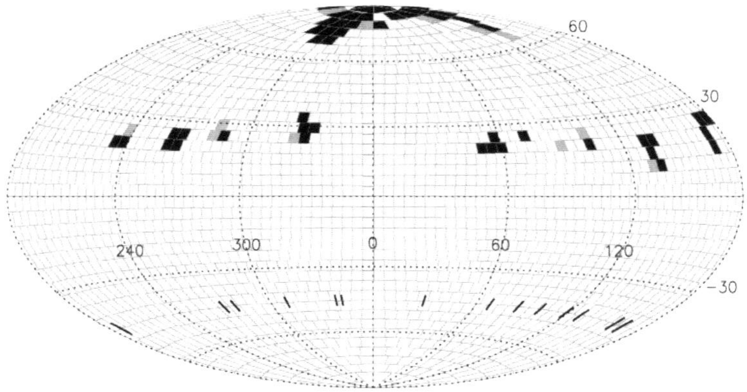

Figure 2: Aithoff projection of our DI based on the 2013 (upper panel) and 2015 (lower panel) NARVAL spectra. The sub-observer longitude at phase zero is the meridian at the centre, marked by zero. Rotation phases sampled with spectra are indicated by marks on the southern hemisphere. Rotation progresses with decreasing surface longitude, i.e. from left to right in this projection. As discussed in our main journal article, we did not use the complete 2015 spectra as input to the CLDI, because intrinsic spot evolution during less than one rotation inhibited the construction of a DI.

4. WHAT DO WE LEARN ABOUT OU AND ?

Strictly speaking, our CLDI Doppler images are maps of photospheric brightness in a narrow subinterval of the visible spectrum. However, in the context of 'cool' stars, like OU And, the CLDI maps can be interpreted as good proxies of temperature maps, though they lack a temperature calibration (see e.g. Wolter et. al., 2005).

Keeping that in mind, our (proxy) temperature maps of OU And for 2008 and 2013 are in a good agreement with the magnetic maps published in BEAL16, which were taken strictly contemporaneously. Our temperature maps confirm the reality of a persistent and large polar spot on OU And, as already found by BEAL16. Furthermore, like the magnetic maps, our temperature maps show smaller cool spots closer to the equator. Unfortunately, we cannot trace the evolution and precise location of these spots on the basis of of our dataset.

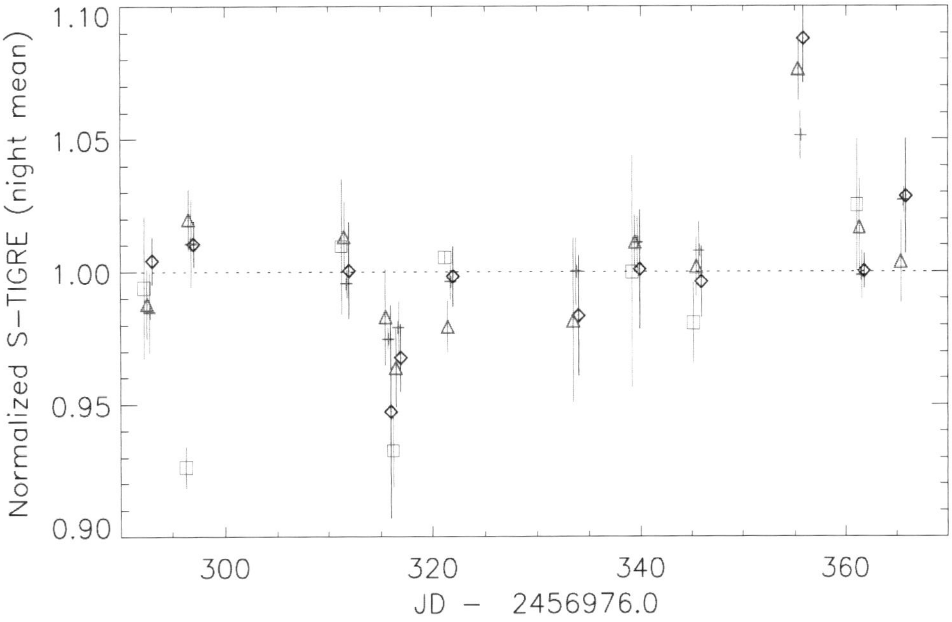

Figure 3: All activity indices measured in our densely time-sampled TIGRE spectra in September 2015. Here, we only show the robustly estimated mean and 1-sigma standard deviation for each night, which is computed after an outlier rejection from typically ten spectra. All points are slightly shifted along the abscissa for each night for better visibility, the corresponding spectra were taken strictly simultaneously. Different symbols represent different lines: the S-Index is represented by violet squares; the TIGRE indices for the Ca IR are marked by blue plus symbols (8489 A), black diamonds (8542 A) and red triangles (8662 A), respectively. Each line index is normalized to the median of its time series.

Our study of the activity indices spans the period 2008 to 2017. In agreement with BEAL16, we found only modest variations in chromospheric activity, e.g. with an amplitude of a few percent peak-to-peak in terms of the Ca II H&K S-index. In the framework of our study, OU And showed the most pronounced chromospheric variations in 2013; this holds true for all activity indices and the corresponding chromospheric emission line profiles. As already discussed by BEAL16, we presume that OU And is an Ap star descendant that has evolved close to the end of the Hertzsprung gap, still retaining a relatively fast rotation from its main sequence phase.

5. CONCLUSIONS

We performed photospheric Doppler imaging (DI) and studied the chromospheric activity of the late-type single giant OU And, based on high resolution spectroscopic data from the period 2008 to 2015. We compared our temperature maps with magnetic maps of the same star created by Zeeman Doppler imaging (ZDI) and presented by Borisova et al. 2016, referenced as BEAL16 here. Both DI methods and studies find a persistent polar spot and weaker, only marginally resolved, lower latitude features on OU And. Furthermore, we studied the chromospheric activity of OU And, based on the spectra taken with the NARVAL instrument, combining them with spectral time series taken with TIGRE/HEROS. In total, our spectroscopic dataset spans the years 2008 to 2017. We find a weak long-term variation of OU And's chromospheric emission and no clear indications of a rotational modulation.

As in the case of other tentative activity cycle candidates, we believe that further spectroscopic long-term studies of OU And will be required to support of refute the reality of a cycle in this case. Such studies will need to span a decade or more. We are continuing to observe OU And with a coordinated use of our telescope facilities in Bulgaria, using the echelle spectrograph ESPERO, and in Mexico, using the HEROS spectrograph.

Acknowledgements

UW acknowledges funding by DLR, project 50OR1701; AB and RK-A acknowledge financial support in the framework of DNTS Germania 01/10. We make use of data obtained with the NARVAL spectrograph (OPTICON programs for semesters 2008B and 2015A), funded by project BG051PO001-3.3.06-0047 and financed by the EU, ESF and Republic of Bulgaria. Based on observations taken with the TIGRE telescope, located at La Luz, Mexico. TIGRE is a joint collaboration of the Hamburger Sternwarte, the University of Guanajuato and the University of Liege.

The lead author of this article, AB, died unexpectedly in December 2017. Ana initiated this study and we have completed it along the lines set out by her and UW several years ago. We mourn the loss of an esteemed colleague and a great person. May Ana rest in peace.

References

Ayres, T. R., Hodges-Kluck, E., Brown, A.: 2007, *ApJS*, **171**,304.

Borisova, A., Auriere, M., Petit, P., Konstantinova-Antova, R., Charbonnel, C., Drake, N. A.: 2016, *A&A*, **591,** A57.

Bruls, J. H. M. J., Solanki, S. K., Schüssler, M.: 1998, *A&A*, **336**, 231.

Cowley, A. P., Bidelman, W. P.: 1979, *PASP*, **91**, 83.

Gray, R. O., Napier, M. G., Winkler, L. I.: 2001, *AJ,* **121**, 2148.

Gondoin, P.: 2003, *A&A*, **409**, 263.

Wolter, U., Schmitt, J. H. M. M., van Wyk, F.: 2005, *A&A*, **435**, 261.

Wolter, U., Robrade, J., Schmitt, J. H. M. M., Ness, J.-U.: 2008, *A&A*, **478**, L11.

Proceedings of the XI Bulgarian-Serbian Astronomical Conference
(XI BSAC) Belogradchik, Bulgaria, May 14-18, 2018
Editors: Milcho K. Tsvetkov, Milan S. Dimitrijević and Momchil Dechev
Publ. Astron. Soc. "Rudjer Bošković" No 18, 2018, 59-66

NEW LINEAR SOLUTIONS FOR 13 DOUBLE STARS

ZORICA CVETKOVIĆ, RADE PAVLOVIĆ,
GORAN DAMLJANOVIĆ and MILJANA D. JOVANOVIĆ

Astronomical Observatory, Volgina 7, 11060 Belgrade, Serbia
E-mail: zorica@aob.rs, rpavlovic@aob.rs, gdamljanovic@aob.rs, miljana@aob.rs

Abstract. We present 13 linear solutions of double stars that have been calculated for the first time from 2015 to 2017. For the purpose of calculating the linear solutions, we also use our measurements of the position angle and angular separation obtained from CCD frames performed at NAOR and ASV. Two linear elements, V_x and V_y , are used to calculate the velocity V of relative motion of the secondary with respect to the primary. Also, we can calculate the relative proper motion μ_{rel} for these 13 double stars by using the components of the proper motion, $\mu_\alpha \cos\delta$ and μ_δ, for primary A and secondary B. The components of the proper motion are taken from two sources, WDS catalog and Gaia DR2. Then, we can compare the proper motion μ_{rel} with the values of the relative velocity V for all 13 components in linear solutions. Excellent agreement between them was found when we used the proper motions from the Gaia DR2 catalog to calculate μ_{rel}.

1. INTRODUCTION

Visual double stars have been studied at the Astronomical Observatory in Belgrade (AOB) for almost 70 years, i.e. from late 1950. These objects had been observed till the end of the XX century with the Zeiss refractor (65/1055 cm). The micrometric measuring technique had been used and in this way from AOB more than 260 new pairs had been discovered. After this, for several years the observations were performed with CCD camera ST6 attached to this telescope. In the beginning of this century observations of double stars from AOB ceased, to be continued from Bulgaria with the 2 m telescope of the National Astronomical Observatory at Rozhen (NAOR). CCD observations with VersArray 1300B camera were performed. From 2004 to 2007 the cooperation between Serbian and Bulgarian astronomers in the field of visual double star observations occurred within the framework of a regional project sponsored by UNESCO-ROSTE. The results have been published in (Pavlović et al. 2005, Cvetković et al. 2006, 2007).

From the Serbian side the participants were: Georgije Popović, Dragomir Olević, Rade Pavlović and Zorica Cvetković, from the Bulgarian one Anton Strigachev.

From 2007 till nowadays the observations of visual double stars have been continued with the 2 m telescope of NAOR, first thanks to the understanding of the Bulgarian colleagues (2009–2011), and then in the framework of joint triennial projects between the Serbian Academy of Sciences and Arts and Bulgarian Academy of Sciences for the periods 2012–2014, 2014–2016 and 2017–2019. The results have been published in (Cvetković et al. 2010, 2011, 2012). The participants from the Serbian side have been: Zorica Cvetković, Rade Pavlović and Goran Damljanović all the time, then Slobodan Ninković (2012–2016), Milan Stojanović (2012–2014) and Miljana Jovanović (from 2017). The Bulgarian participants have been: Svetlana Boeva all the time, Alexander Antov, Renada Konstantinova-Antova, Rumen Bogdanovski and Borislav Spassov (2012–2014), Georgi Latev and Maya Belcheva (from 2014) and Yanko Nikolov (from 2017).

Immediately after finishing the building of the new Astronomical Station on the mountain of Vidojevica (ASV) in southern Serbia, in the middle of 2011 CCD observations of visual double stars with a 0.6 m telescope were initiated. The results have been published in (Pavlović et al. 2013; Cvetković et al. 2015, 2016, 2017). When a larger telescope named "Milankovic" with the diameter of the primary mirror equal to 1.4 m was procured in 2016, CCD observations of these objects have been also performed by using it. More details about ASV and its telescopes can be found on the websites http://vidojevica.aob.rs/ and http://belissima.aob.rs/. Two CCD cameras have been utilized at ASV, Apogee Alta U42 and SBIG ST-10ME.

The CCD frames were measured using AIP4WIN software (Berry and Burnell, 2002) and the IRAF[1] package. The observables are two relative coordinates, angular separation between components A (primary) and B (secondary), or separation only, ρ and position angle θ, for the moment of observation t. The task is to establish which double stars are gravitationally bound, or orbital pairs, unlike those not gravitationally bound, i.e. are optical pairs, with close projections onto the tangential plane. Further the orbital or linear elements should be calculated, in order to determine stellar masses, dynamical parallaxes and other parameters.

The linear elements are: X_0 and Y_0 – coordinates of the closest approach of the secondary star relative to the primary; V_X and V_Y – components of the velocity of relative motion of the secondary relative to the primary; T_0 – epoch of the closest approach; ρ_0 – the closest relative separation and θ_0 – position angle of the closest approach. A more detailed explanation concerning the linear elements, figures and the formulae for their obtaining is given in another paper (Pavlović et al. 2018).

[1] http://iraf.noao.edu

2. LINEAR SOLUTIONS

From 2012 we have calculated for the first time the linear elements for a total of 27 double stars. Out of them linear solutions for 13 double stars were obtained between 2015-2017, which are presented here. Their identifications in the Washington Double Stars Catalog (WDS) and discoverer designation are: WDS 01057+3304=MLB 444, WDS 02516+4803=HJ 2160AB, WDS 03342+4837=BU 787AB, WDS 04312+5858=STI 2051AB, WDS 04556+1653=HJ 3263, WDS 05492+2941=BRT 2521, WDS 06092+6424=MLB 259, WDS 07106+1543=J 703, WDS 08503+0125=J 74, WDS 09388+0242=J 78, WDS 17046+3900=HJ 2804AB, WDS 18269+2950=HJ 1325 and WDS 19500+0637=J 1336AB. For the purpose of calculating the linear solutions, in addition to the measurements taken from the database, we also use our measurements of the position angle and angular separation from CCD frames obtained at NAOR and ASV. The linear elements are given in Table 1. They have been already announced so that they are available in the Catalog of Rectilinear Elements (http://www.astro.gsu.edu/wds/lin1.html).

Table 1: Linear elements for 13 double stars.

WDS Designation	Discoverer Designation	X_0 ($''$)	X_A ($''$/year)	Y_0 ($''$)	Y_A ($''$/year)	T_0 (year)	ρ_0 ($''$)	θ_0 (°)
01057+3304	MLB 444	7.184849	−0.029876	−2.073079	−0.103544	1948.782	7.478	73.91
02516+4803	HJ 2160AB	−5.833506	−0.013444	3.726437	−0.021046	1743.980	6.922	237.43
03342+4837	BU 787AB	−1.744095	−0.020860	1.467897	−0.024785	1878.073	2.258	229.91
04312+5858	STI2051AB	5.579910	0.042772	2.330326	−0.102417	1939.958	6.047	112.67
04556+1653	HJ 3263	−3.459261	0.039617	−2.136124	−0.064157	1834.786	4.066	301.70
05492+2941	BRT2521	−1.248724	−0.076498	2.026629	−0.047135	1940.824	2.380	211.64
06092+6424	MLB 259	−3.980805	0.013260	1.576882	0.033474	1910.178	4.282	248.39
07106+1543	J 703	0.415380	−0.082381	−1.745080	−0.019609	1892.897	1.794	3.39
08503+0125	J 74	2.191639	0.034511	−1.685230	0.044882	1857.508	2.765	52.44
09388+0242	J 78	0.562050	0.067989	3.036883	−0.012583	1948.224	3.088	169.51
17046+3900	HJ 2804AB	−2.604337	0.083382	4.882575	0.044476	2049.773	5.534	208.08
18269+2950	HJ 1325	−5.030179	0.009846	−4.126053	−0.012003	2001.494	6.506	309.36
19500+0637	J 1336AB	−0.370453	0.047707	−2.036395	−0.008679	1904.829	2.070	349.69

Two linear elements, V_x and V_y , are used to calculate the velocity V of relative motion of the secondary with respect to the primary. Also, we can calculate the relative proper motion μ_{rel} for these 13 double stars by using the components of the proper motion, $\mu_\alpha \cos\delta$ and μ_δ, for primary A and secondary B. The components of the proper motion are taken from two sources, the WDS catalogue (http://www.usno.navy.mil/USNO/astrometry/optical-IR-prod/wds) and Gaia DR2 (http://vizier.u-strasbg.fr/viz-bin/VizieR-3?-source=I/345/gaia2). They are given in Table 2. The Gaia DR2 catalogue is available from April 2018. It contains high accuracy data and there one can find parallaxes for both components of all the 13 pairs from our sample. The Gaia DR2 parallaxes are also given in Table 2.

Table 2: The components of the proper motion, $\mu_\alpha\cos\delta$ and μ_δ, for primary A and secondary B taken from WDS and Gaia DR2 and parallaxes π_A and π_B from Gaia DR2.

			WDS catalogue				Gaia DR2 catalogue			
Pair	π_A (mas)	π_B (mas)	$\mu_{\alpha A}\cos\delta$ (mas/yr)	$\mu_{\delta A}$ (mas/yr)	$\mu_{\alpha B}\cos\delta$ (mas/yr)	$\mu_{\delta B}$ (mas/yr)	$\mu_{\alpha A}\cos\delta$ (mas/yr)	$\mu_{\delta A}$ (mas/yr)	$\mu_{\alpha B}\cos\delta$ (mas/yr)	$\mu_{\delta B}$ (mas/yr)
MLB 444	3.5453	3.4627	−18.8	−102.8	87.7	23.7	11.900	−88.981	−14.926	9.472
HJ 2160AB	1.0469	2.3594	8.8	−7.3	−8.0	13.9	9.259	−7.452	−1.936	14.074
BU 787AB	5.8057	1.6200	21.4	−26.5	22.1	17.8	22.419	−26.732	−4.788	1.288
STI2051AB	180.4215	181.2815	1300.2	−2049.0	1335.6	−1962.6	1303.270	−2043.847	1335.042	−1947.632
HJ 3263	2.7324	0.8136	−44.3	−56.5	5.7	5.9	−48.778	−61.537	2.802	−2.488
BRT2521	7.6388	0.9987	92.0	−55.8	−14.9	−41.9	67.125	−56.269	−7.493	−5.842
MLB 259	2.9454	1.3870	4.6	17.8	−15.0	−33.2	−4.564	13.385	11.251	−21.119
J 703	2.7841	12.1025	8.9	−16.9	−23.0	15.8	−12.119	−5.208	−93.538	14.491
J 74	4.4225	1.6365	−61.7	55.8	40.1	−15.5	−40.064	38.112	−7.610	−5.594
J 78	7.9151	5.5372	−76.0	0.0	−30.9	−58.1	−94.165	−30.793	−22.601	−17.829
HJ 2804AB	5.5242	3.4322	−51.7	2.0	25.3	−44.2	−53.121	1.738	30.945	−43.992
HJ 1325	0.3623	0.4096	−1.3	−11.6	−21.0	12.1	−3.200	−11.248	5.956	0.604
J 1336AB	1.0174	4.5890	−11.4	−18.0	28.8	8.7	0.246	−6.148	44.409	−6.481

The relative proper motion is designated as $\mu_{rel}(W)$ when we used the proper motions from the WDS catalogue, and as $\mu_{rel}(G)$ when we used the proper motions from Gaia DR2 catalogue. The relative proper motions $\mu_{rel}(W)$ and $\mu_{rel}(G)$ and the velocity V of relative motion of the secondary with respect to the primary are given in Table 3.

Then, we compared μ_{rel} with V for all 13 components in the linear solutions. As can be seen from Table 3, the values of V are in excellent agreement with $\mu_{rel}(G)$ for all 13 pairs and with $\mu_{rel}(W)$ the agreement is good for eight pairs, HJ 2160, BU 787, STI 2051, HJ 3263, BRT 2521, J 78, HJ 2804 and J 1336. There is a disagreement in V and $\mu_{rel}(W)$ for five pairs: MLB 444, MLB 259, J 703, J 74 and HJ 1325. The reason of disagreement is unreliable proper motion components for the primary and/or secondary in WDS. In the case of these five pairs, as can be seen from Table 2, $\mu_\alpha\cos\delta$ and μ_δ from WDS and Gaia DR2 are significantly different. An agreement between μ_{rel} and V is an argument in favor that the pairs are not gravitationally bound, but instead mutually very distant in space so that only their projections are close in the field of view. Now, the Gaia DR2 catalog is available. In it one can find the parallaxes for both components for all 13 pairs. As can be seen (Table 3), the parallaxes of primary π_A and secondary π_B are different for 11 pairs, i.e. the components are at substantially different heliocentric distances. This is strongly in favor that they are not very long-period physical pairs, but just optical pairs. In the case of two pairs, MLB 444 and STI 2051, the parallaxes of primary π_A and secondary π_B are almost equal to each other, i.e. the components are at the same heliocentric distance. Most likely these two pairs are not bound, but their components have common proper motion. The linear fits for these 13 double stars are presented in Figures 1-3.

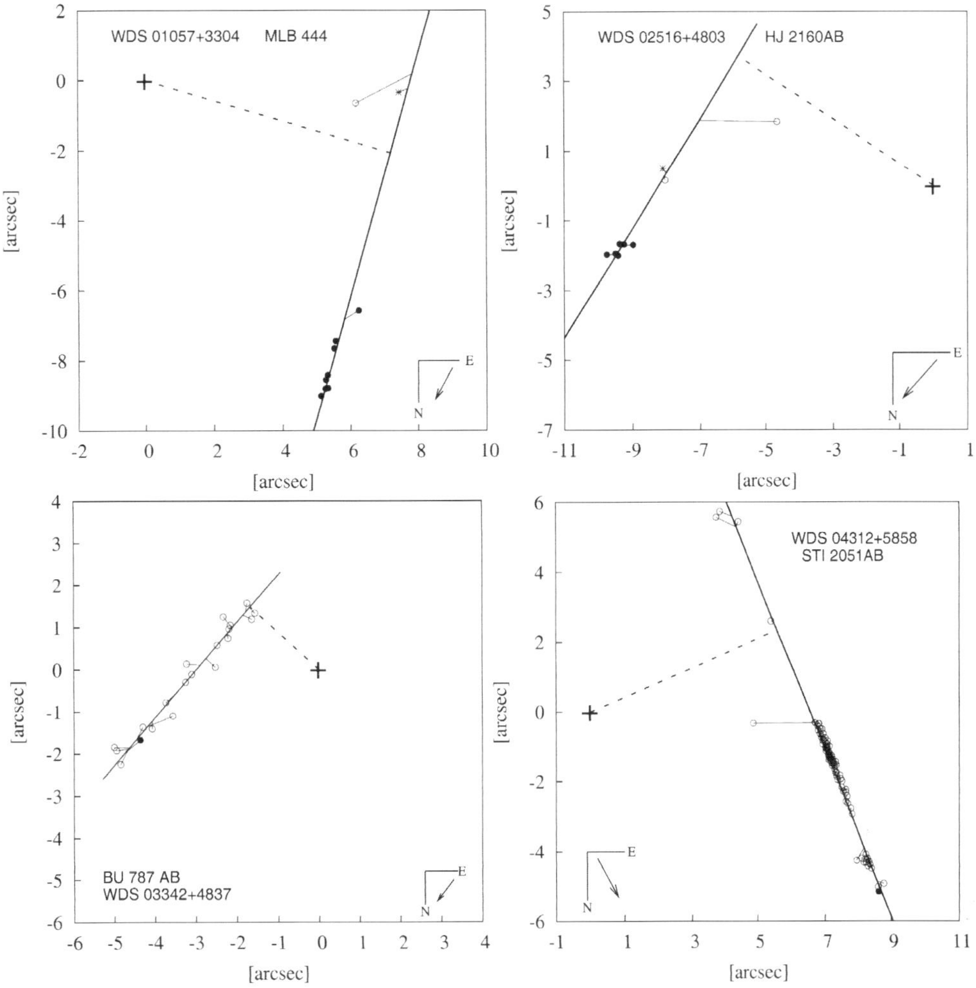

Figure 1: Linear fits for pairs MLB 444, HJ 2160AB, BU 787, STI 2051, HJ 3263 and BRT 2521. The arrow in the lower right corner indicates the direction of the relative motion of the secondary; the dashed perpendicular line from the linear fit to the origin indicates the closest relative separation. The micrometric and photographic observations are represented by open circles and asterisks, respectively. CCD or other two-dimensional electronic imaging observations are denoted by filled circles.

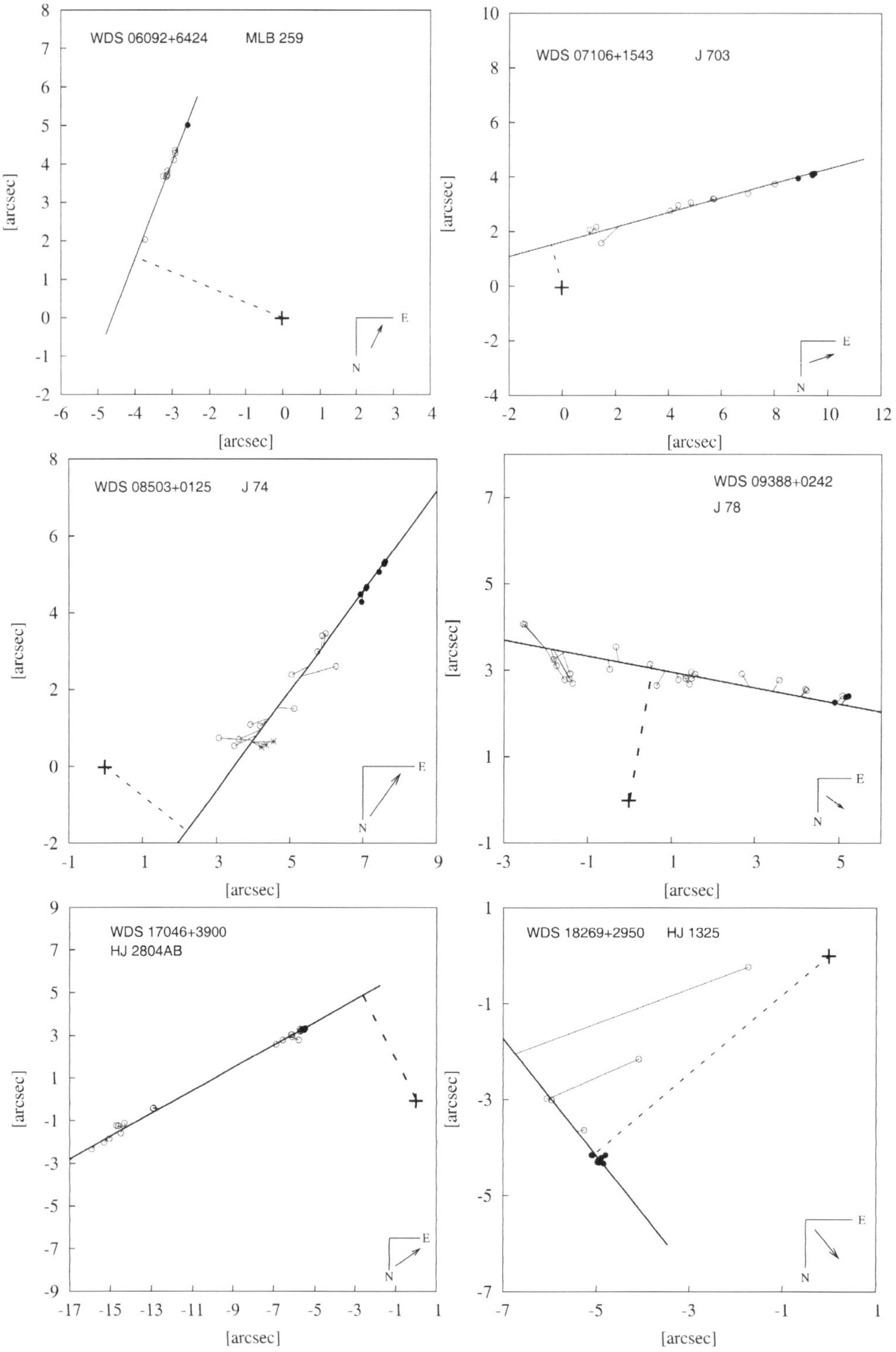

Figure 2: Linear fits for pairs MLB 259, J 703, J 74, J 78, HJ 2804 and HJ 1325. All designations on the plots are the same as in Figure 1.

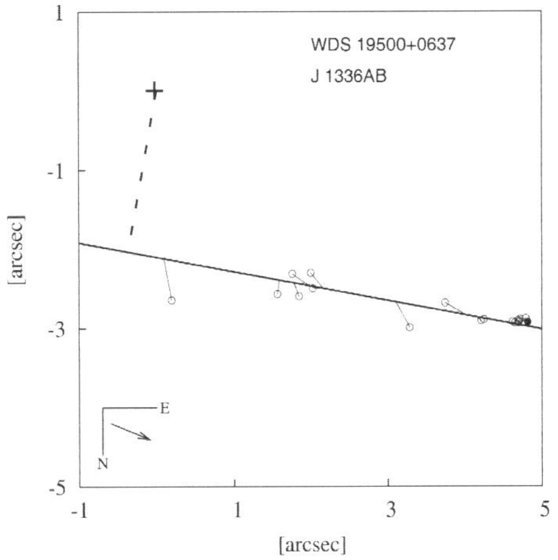

Figure 3: Linear fit for pair J 1336. All designations are the same as in Figure 1.

Table 3: The relative proper motions $\mu_{rel}(W)$ and $\mu_{rel}(G)$ and the velocity V of relative motion of the secondary with respect to the primary.

Pair	$\mu_{rel}(W)$ [″/yr]	$\mu_{rel}(G)$ [″/yr]	V [″/yr]
MLB 444	0.165	0.102	0.108
HJ 2160AB	0.027	0.024	0.025
BU 787AB	0.032	0.039	0.032
STI2051AB	0.093	0.101	0.111
HJ 3263	0.080	0.078	0.075
BRT2521	0.108	0.090	0.090
MLB 259	0.055	0.038	0.036
J 703	0.046	0.084	0.085
J 74	0.124	0.054	0.057
J 78	0.074	0.073	0.069
HJ 2804AB	0.090	0.096	0.094
HJ 1325	0.031	0.015	0.016
J 1336AB	0.048	0.044	0.048

5. CONCLUSIONS

From the comparison of the proper motion μ_{rel} with the values of the velocity V in the linear solutions for 13 pairs we found excellent agreement between them when the proper motions from Gaia DR2 for calculation of $\mu_{rel}(G)$ were used.

The WDS catalog now contains information about the measurements for 142645 double stars, where only for 2775 of them orbits have been calculated, whereas for 1574 double stars there exist linear elements. In the case of the remaining very numerous visual double stars the nature of their motion is to be determined only, to which the Gaia catalog, available from recent times, will contribute significantly. In addition to precise positions, Gaia DR2 provides parallaxes and proper motions for both components for most double stars. Thanks to these parallaxes, in the near future, it will be possible to segregate in WDS gravitationally unbound pairs (components at different distances) from those with Keplerian motion.

Acknowledgments

We gratefully acknowledge the observing grant support from the Institute of Astronomy and Rozhen National Astronomical Observatory, Bulgarian Academy of Sciences. This research has been supported by the Ministry of Education, Science and Technological Development of the Republic of Serbia (Projects No. 176011 "Dynamics and Kinematics of Celestial Bodies and Systems").

References

Berry, R., Burnell, J.: 2002, *The Handbook of Astronomical Image Processing*, Includes AIP4WIN Software (Richmond, VA: Willmann-Bell).

Cvetković, Z., Novaković, B., Strigachev, A., Popović, G. M.: 2006, "CCD measurements of double and multiple stars at NAO Rozhen. II", *Serb. Astron. J.*, **172**, 53-58.

Cvetković, Z., Pavlović, R., Strigachev, A., Novaković, B., Popović, G. M.: 2007, "CCD measurements of double and multiple stars at NAO Rozhen. III", *Serb. Astron. J.*, **174**, 83-88.

Cvetković, Z., Pavlović, R., Boeva, S.: 2010, "CCD measurements of double and multiple stars at NAO Rozhen. IV", *Serb. Astron. J.*, **180**, 103-107.

Cvetković, Z., Pavlović, R., Damljanović, G., Boeva, S.: 2011, "CCD measurements of double and multiple stars at NAO Rozhen. Orbits and linear fits of five pairs", *Astronomical Journal*, **142**, 73.

Cvetković, Z., Pavlović, R., Boeva, S., Damljanović, G.: 2012, "CCD measurements of double and multiple stars at NAO Rozhen", *Bulgarian Astronomical Journal*, **18(1)**, 56-61.

Cvetković, Z., Pavlović, R., Boeva, S.: 2015, "CCD measurements of double and multiple stars at NAO Rozhen and ASV in 2012. Four linear solutions", *Astronomical Journal*, **149**, 150.

Cvetković, Z., Pavlović, R., Boeva, S.: 2016, "CCD measurements of double and multiple stars at NAO Rozhen and ASV in 2013 and 2014. Eight linear solutions", *Astronomical Journal*, **151**, 58.

Cvetković, Z., Pavlović, R., Boeva, S.: 2017, "CCD measurements of double and multiple stars at NAO Rozhen and ASV in 2015", *Astronomical Journal*, **153**, 195.

Pavlović, R., Cvetković, Z., Olević, D., Strigachev, A., Popović, G. M., Novaković, B.: 2005, "CCD measurements of double and multiple stars at NAO Rozhen", *Serb. Astron. J.*, **171**, 49-53.

Pavlović, R., Cvetković, Z., Boeva, S., Vince, O., Stojanović, M.: 2013, "CCD measurements of double and multiple stars at NAO Rozhen and ASV in 2011. Five linear solutions", *Astronomical Journal*, **146**, 52.

Pavlović, R., Cvetković, Z., Boeva, S., Damljanović, G.: 2018, "Determination of orbits of visual binary and linear elements of double stars", *Astronomical and Astrophysical Transactions*, vol. **30**, 3-4 (in press).

Proceedings of the XI Bulgarian-Serbian Astronomical Conference
(XI BSAC) Belogradchik, Bulgaria, May 14-18, 2018
Editors: Milcho K. Tsvetkov, Milan S. Dimitrijević and Momchil Dechev
Publ. Astron. Soc. "Rudjer Bošković" No 18, 2018, 67-74

THE BROKEN TITIUS-BODE LOW FOR THE SOLAR PLANETS AND FOR THE SATELLITE SYSTEMS OF THE JOVIAN PLANETS

TSVETAN B. GEORGIEV

Institute of Astronomy and National Astronomical observatory, Bulgarian Academy of Sciences, 72 Tsarigradsko Chausse blvd., Sofia 1784, Bulgaria
E-mail: tsgeorg@astro.bas.bg

Abstract. In this paper we build and compare TBL models for the planets and dwarf planets in the Solar System, as well as for regular and dwarf satellites of the Jovian planets. The TBL models of the planetary system and satellite systems of the Jovian planets seem to be broken or dualistic. While the solar planets and the regular satellites of the Jovian planets obey TB relations within relative standard error about 10 %, the small inner satellites of the Jovian planets obey their own TB relations with low TBL gradient and standard errors about 5 %. In the cases of Jupiter, Saturn and Neptune the bounds of the planetary rings, marked along the ordinate axes of log *P* (orbital period) envelop the rotational periods of the planet under TBL No. 0. (at Saturn – together with the missing satellite No.1). The rings of Jupiter, Saturn and Neptune correspond to the missing small satellites under No.0 (at Saturn – together with the missing satellite under No.1). However, the rings of Uranus is placed above the known inner satellites. The rings of Uranus may be associated with a missing satellite under No.6

1. INTRODUCTION

The Titius-Bode low (TBL) is a generalization of the Titius-Bode rule found in the last third of 18th century (Goldreich 1965, Dermott 1968, Nietto 1972). The TBL contains 4 basic topics:

First. In every system of orbiting bodies (solar planets, planet satellites, exoplanet systems) the orbital size (or period) grows up with the distance from the gravitational center near-commensurability, following a power function on the number of the orbital size/period. The conventional model of the TBL for the periods *P* today is

$$(1) \qquad P_n = P_0.P_c^{\,n} \quad \text{or} \quad \log P_n = \log P_0 + \log P_c.n,.$$

Here $n = 1, 2, \ldots, N$ is the period number, P_n is the nth period, P_0 and P_c are constants. P_0 is the scale factor (amplitude coefficient) with meaning of period under number 0 and P_c is the power factor of the regularity (near-commensurability) of the orbital periods. The TBL may be written by exponential function and natural logarithms too (Poveda & Lara 2008, Panov 2009). Other models of the distributions of the planetary distances exists too (Carry 1988, Kotliarov 2008, Ignatovich 2014, Ignatovich 2018).

Second. Every orbiting system has its own constants P_0 and P_c, which might be derived empirically. The best way is the deriving of the regression line of the logarithmic form of the TBL.

Third. In principle the TBL concerns the "regular" objects in the system – relatively massive bodies with almost circular and almost complanar orbits.

The TBL model (Eq.1) has been applied firstly occur unique and conclusive. So,this problem has been settled for decades past. Recently on the solar planets and on the regular satellites of Jupiter (4 satellites), Saturn (7) and Uranus (5) by Dermott (1968). The TBL models TBL models for the regular satellites of Neptune (3) and Pluto (3) were derived on contemporary data by Georgiev (2016).

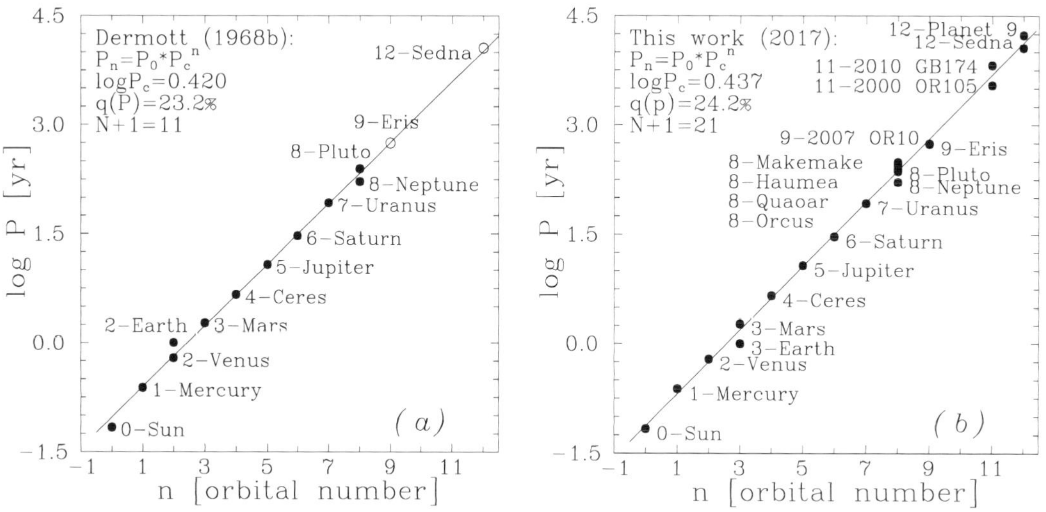

Figure 1: TBL regression models including the rotational period of the Sun under No. 0. (*a*) Classic TB relation (Dermott, 1968). (*b*) Full TB relation, based on contemporary data (Georgiev 2017). The standard errors are large, about 23 % and 24 %, respectively. Note that the Earth seems to be misplaced in both cases.

Regular bodies among the solar planets are 4 Jovian planets and (not obligatory) 4 Terrestrial planets, but Dermott (1968) was used 10 bodies – 8 planets plus Ceres and Pluto (Fig.1,*a*). The upper mentioned regular planet satellites were used in the papers of Georgiev (2016, 1917). The full contemporary model includes 8 planets and 12 dwarf planets with computer defined optimal numbering of the orbits (Fig.1,*b*).In the presented other TB relation of the solar planets are shown in Fig.2.

In the Solar System the TBL reveals one important particularity. The rotational period of the central body, under TBL No. 0, supports approximately the TBL model and this fact is used in the papers of Dermott (1968) и Georgiev (2016, 2017). However, such support is not observing even once among all 8 exoplanet systems with known rotational period of the star (Georgiev 2018). The reason for this particularityis not clear. Though, in the presented paper the rotational periods of the central bodies are not used in the TB regressions (Fig.1, left panel of Fig.2, Fig. 3, Fig.4).

The TBL is not explained conventionally yet (Hayes & Tremaine 1998, Lynch 2003), but it occur fulfilled for all, more than 200 exoplanet systems with known at least 3 exoplanets. About 100 holes in the orbital sequences of these systems, corresponding by presumption to stable orbits, are known (Bovaird & Lineweaver 2013, Bovaird et al. 2015). Georgiev (2018) regarded 30 orbital systems (17 exoplanet systems, 4 versions of the Solar systems, 5 systems of regular planetary satellites and 4 systems of small inner satellites).

Today any example of not-performance of the TBL would be regarded as unexpected and important news. For this reason the interest in the TBL is increasing. So, the TBL is considered as a fundamental natural law, which realizations, especially in the Solar System, need new additional attention.

One basic question is: Whether the constants of the TBL correlate with other, more fundamental parameters of the system, such as the mass of the central body and the total mass of the orbiting bodies? Georgiev (2016, 2017) regards 6 realizations of the TBL in the Solar System and 17 such among the exoplanet systems. The answer is "Yes". The correlations might be useful for the conventional explaining of the TBL.

Another basic question is: Whether the TBL model, derived through the regular bodies, is valid for small distant bodies? In the Solar system numerous (but not all) dwarf planets with elliptic and arbitrary oriented orbits, even with retrograde orbits, follow well the TBL, derived by the regular bodies (Dermott 1968, Georgiev 2016). This interesting circumstance hampers additionally the understanding of the TBL.

Third question, which is regarding in the presented paper is: Whether the TBL model derived through the regular satellites, is valid for the small inner satellites of the Jovian planets? The results in the presented paper claim "No". The inner small satellites obey their own TBL model. Thus the details about the TBL become more complicated, as follows.

2. THE BROJEN TITIUS-BODE RELATIONS

The left panel of Fig.2 shows separate TBL models (solid lines) for 2 group planets (dots) with implicit numbering, derived without participation of the solar rotational period. The TBL model for all 8 planets is shown too (dashed line). The positions of the rotational period of the Sun and orbital periods of 3 dwarf planets are shown by circles. The relative standard deviations of the models are 12 %, 15

% and 21 %, respectively. Thus, the TBL model for the regular solar planets seems to be broken in the region of the main asteroid belt.

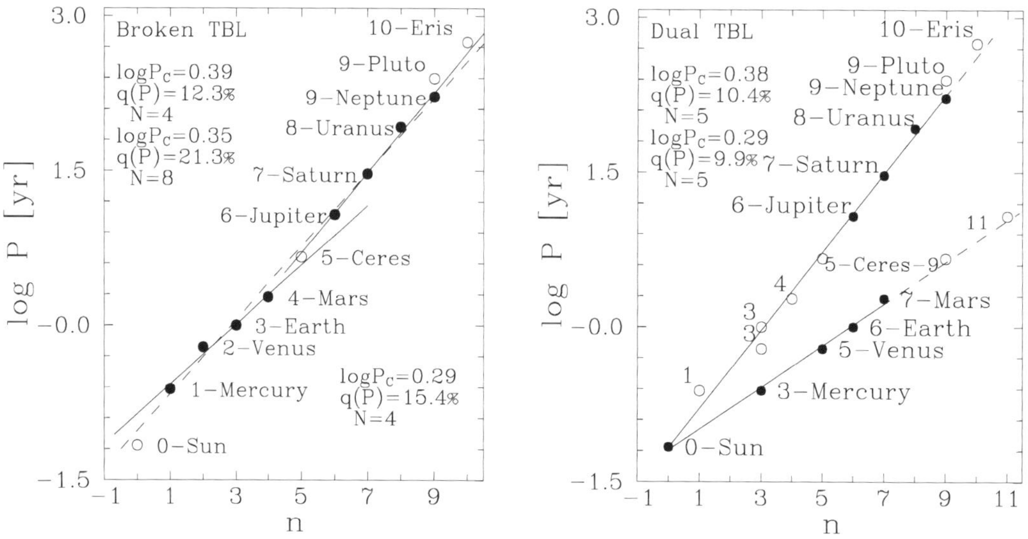

Figure 2: TBL regression models for the regular solar planets (dots) and the positions of other objects (circles). Abscissa axes – numbers of the period (invariant in respect to additive number); Ordinate axes – logarithms of the periods in Earth years. Here log P_c is the TBL gradient, $q(P)$ is the relative standard deviation of the model, N is the number of used periods. (See the text.)

The right panel of Fig.2 shows separate TBL models (solid lines) for the Jovian and Terrestrial planets (dots), derived with participation of the solar rotational period under No.0 (dot). Such kind of diagram is known so far (Rubčic̆ & Rubčic̆ 1995). Here the optimal numbers of the planets are derived by suitable computer method (Georgiev, 2018). The positions of the orbital periods of other objects are shown by circles. The relative standard deviations of both models are about 10 %. Thus, the TBL model for the regular solar planets seems to be dual.

The gradient of the TBL model in the right panel, derived by the Jovian planets (which are just the regular objects in the Solar System) is close to the "standard" one, with log $P_c \approx 0.4$. According to this model and other similar models (Dermott 1968, Georgiev 2016) Venus and the Earth receive one number, here No.3. However, here Venus seems more irrelevant than the Earth.

Though, the gradient, derived by the Terrestrial planets is significantly less, log $P_c \approx 0.3$. (The respective coefficients of near-commensurability of the periods are $P_c \approx 2.5$ and $P_c \approx 1.9$.) Now, in the TBL model for the Terrestrial planets Venus and the Earth receive different numbers, 5 and 6, but dome numbers rest empty.

Figures 3 and 4 show the broken TBL models for the satellite systems of the Jovian planets. There the top placed lines show the down part of the TBL models for the regular satellites (4 of Jupiter, 7 of Saturn, 5 of Uranus and 3 of Neptune). The rotational periods of the planets are not used for the models. The TBL gradients are about 0.3 and the relative standard deviations of these models are about 10 % (Georgiev, 2016, 2018).

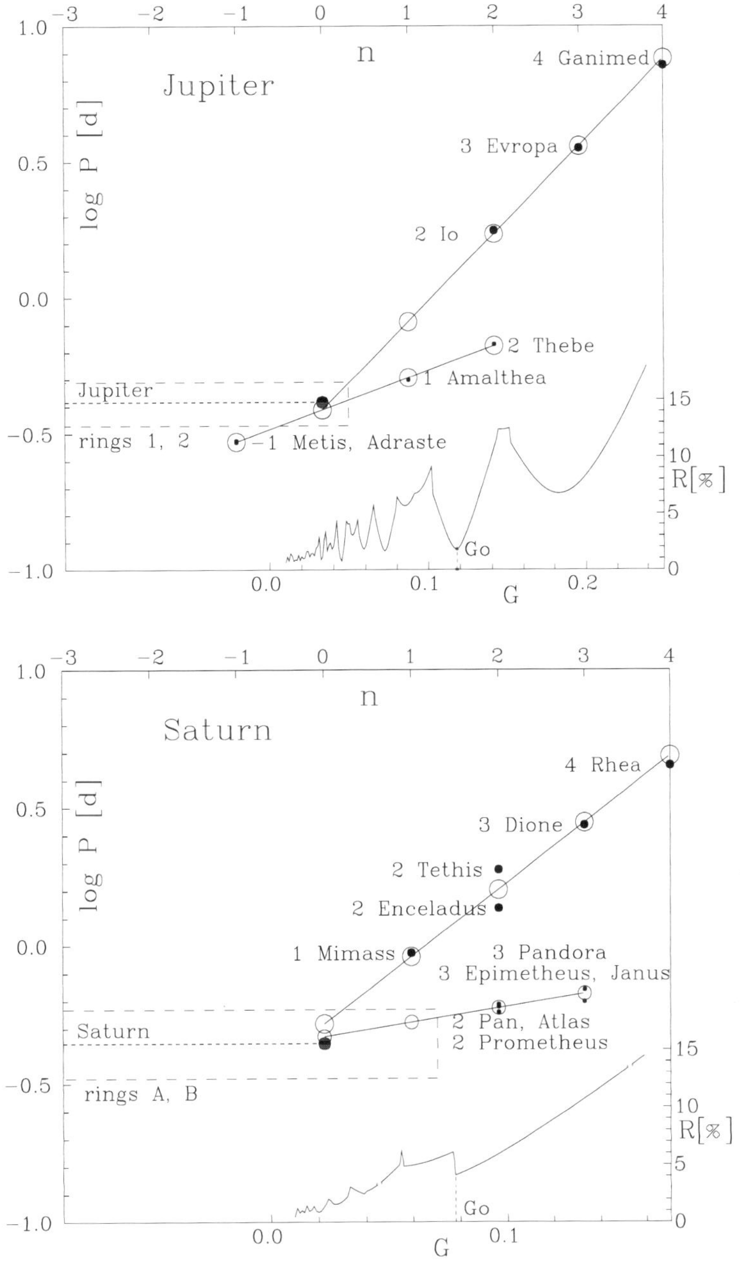

Figure 3: TB relations in the he systems of Jupiter and Saturn. TBL regression modes (solid lines) for the regular planet satellites (large dots), small inner satellites (small dots) and predicted positions of available or missing satellites (circles). The error curves, used for deriving of the optimal gradients G_0 and respective optimal numbering of the inner satellites (Georgiev 2018) are implanted in the right bottom corners. Dashed lines show the approximate bounds of the planetary rings along the axes log P.

The down placed lines in Fig.3 and 4 show the TBL models for the inner satellites with diameters 15-190 km (4 at Jupiter, 6 at Saturn, 10 at Uranus and 6 at Neptune). The places of the minimums of the error curves show the optimal gradients of these TBL models, G_0 = log P_c, about 0.1, The levels of the minimums along the down marked right ordinate axis show that the relative standard deviations of the TBL models for the inner satellites are about 5 %.

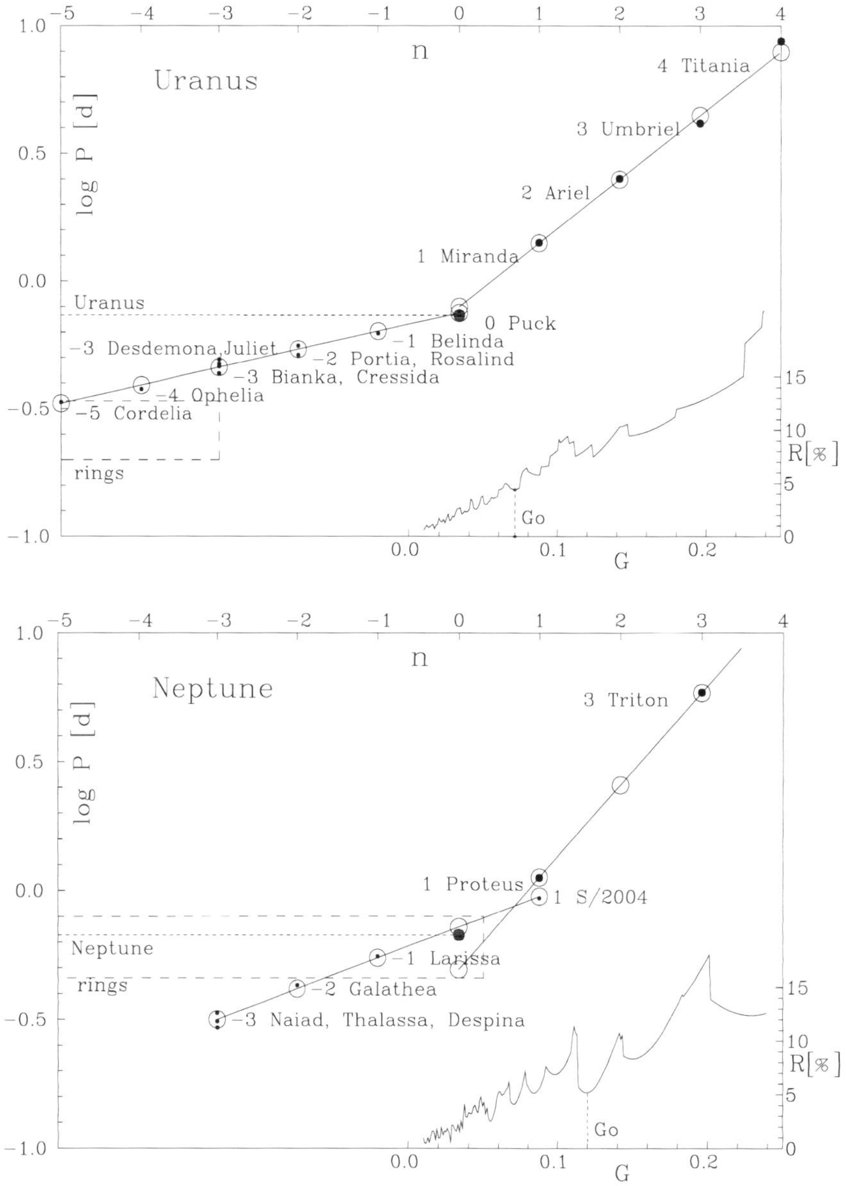

Figure 4: TB relations in the he systems of Uranus and Neptune. See Fig.3.

In Fig.3 and 4 both TBL models of the satellites are shifted horizontally under condition to have common position with the rotational period of the planet under No.0. By this reason some inner satellites receive negative numbers. (The TBL is invariant in respect to addition of a positive or negative number.)

3. RESULTS

The basic results of this work follow.

1. In the cases of Jupiter, Saturn and Uranus the TBL modes, build by the regular satellites predict the rotational period of the planet within error of ± 10 %. However, from this point of view the expected rotational period of the Neptune occur with about 50 % shorter, For the Sun, by the 8 regular planets or for the Jovian planets only, it is longer with about 50 %.

2. The small inner satellites of the Jovian planets obey their own TBL models with low TBL gradient, about 0.1, and small relative standard deviation, about 5 % (Fig.3, 4).

3. The TBL models of the satellite systems of the Jovian planets (Fig.3, 4) seem to be broken like the TBL models for 8 regular planets (Fig.2, left panel).

4. In the cases of Jupiter, Saturn and Neptune the bounds of the planetary rings, marked along the ordinate axes of log *P* (Fig 3 and 4) envelop the rotational periods of the planets and missing small satellites under No. 0. (at Saturn – together with the missing satellite under No.1). However, the rings of Uranus is placed above the known inner satellites. The rings of Uranus may be associated with missing satellite under No.6

References

Bovaird, T., Lineweaver, C. H.: 2013, "Exoplanet predictions based on the generalized Titius–Bode relation", *Mon. Not. Roy. Astron. Soc.,* **435 (2)**, 1126-1138 (arXiv:1304.3341).

Bovaird, T, Lineweaver, C. H., Jakobsen, S. K.: 2015, "Using the inclinations of Kepler systems to prioritize new Titius-Bode-Based exoplanet predictions", *Mon. Not. Roy. Astron. Soc.*, **448**, 3608-3627.

Carrey, S. W.: 1988, *Theories of the Earth and Uuniverse. History of dogma in the Earth sciences*, Stanford University Press, Stanford, CA.

Dermott, S.: 1968, "On the origin of commensurabilities in the solar system - II. The orbital period relation". *Mon. Not. Roy. Astron. Soc.*, **141**, 363-376.

Georgiev, Ts. B.: 2016, "Titius-Bode law in the Solar System. Dependence of the regularity parameter on the central body mass". *Bulg. Astron. J.* **25**, 3 – 18; (http://www.astro.bas.bg/AIJ/).

Georgiev, Ts. B.: 2017, Titius-Bode law and fundamental parameters of the orbital systems in the Solar System", *Publ. Astron. Soc. Bulg.,* **2017**, 1 – 33 (in Bulgarian) (http://astro.shu-bg.net/pasb/).

Georgiev, Ts. B.: 2018, "Correlations of parameters of Titius-Bode law with parameters of exoplanet and satellite systems", *Bulg. Astron. J.*, **29**, 1-29; (http://www.astro.bas.bg/AIJ/).

Goldreich, P.: 1965, "An explanation of the frequent occurrence of commensurable mean motions in the solar system", *Mon. Not. Roy. Astron. Soc.,* **130**, 159-181.

Hayes, W., Tremaine, S.: 1998, "Fitting Selected Random Planetary Systems to Titius-Bode Laws", *Icarus,* **135**, 549-557.

Ignatovich, S.: 2014, "Zakoni planetarnih rastojania na Slovenskom Jugu", *PASRB (Publications of Astronomical Society "Rudjer Bošković")*, **13**, 525-538 (In Serbo-Croatian)

Ignatovich, S.: 2018, "Zakoni planetarnih rastojania na Slovenskom Jugu. II Dio: Novi razvoj.", *PASRB (Publications of Astronomical Society "Rudjer Bošković")*, **17**, 399-413 (In Serbo-Croatian).

Kotliarov, I.: 2008, "A structural low of planetary systems", *Mon. Not. Roy. Astron. Soc.*, **390**, 1411-1412.

Lynch, P.: 2003, "On the significance of the Titius-Bode law for the distribution of the planets", *Mon. Not. Roy. Astron. Soc.*, **341**, 1174-1178.

Nieto, M. M.: 1972, *The Titius-Bode Law of Planetary Distances: its History and Theory.* Pergamon Press Oxford.

Panov, K. P.: 2009, "The orbital distances law in planetary systems", *Open Journal of Astron. Ap*, **2**, 90-94.

Poveda, A, Lara, P.: 2008, "The Exo-Planetary System 55 Canceri and the Titius-Bode Low:" *Rev. Mex. A. A.* **44**, 243–246.

Rubčić, A., Rubčić, J.: 1995, "Stability of gravitationally bound many body systems", Fizika, **B4**, 11-28.

Proceedings of the XI Bulgarian-Serbian Astronomical Conference (XI BSAC) Belogradchik, Bulgaria, May 14-18, 2018
Editors: Milcho K. Tsvetkov, Milan S. Dimitrijević and Momchil Dechev
Publ. Astron. Soc. "Rudjer Bošković" No 18, 2018, 75-86

THE CURRENT STATE OF THE ODESSA COLLECTION OF ASTROPHOTONEGATIVES

SVITLANA KASHUBA[1], MILCHO TSVETKOV[2,3]
and NATALYA BAZYEY[1]

[1]*Astronomical Observatory of Odessa National University, 65014 Odessa, Ukraine*
[2]*Astronomical Institute with National Astronomical Observatory, Bulgarian Academy of Sciences, 1784 Sofia, Bulgaria*
[3]*Institute of Mathematics and Informatics, Bulgarian Academy Sciences, 1113 Sofia, Bulgaria*
E-mail: sv-k@onu.edu.ua, milcho.tsvetkov@gmail.com, n.bazyey@onu.edu.ua

Abstract. We present the status, tasks and opportunities of the Astronomical observatory of the Mechnikov Odessa National University astrophotonegatives collection. It contains one of the most valuable plates archives as it is listed in the Wide-Field Plate Database (wfpdb.org) with more than 100,000 plates – the second plate collection in Europe after the Sonnenberg plate collection with 250,000 plates and the third in the world after the Harvard plate collection with 500,000 plates.

The Odessa plate collection provides an opportunity to follow a brightness history of selected sky objects during an almost 100-year period. We describe the plans of the observatory of salvaging this important world astronomical heritage and the first steps of digitization of an important part of the plate collection. The inventory and incorporation of the metadata from the logbooks in the computer readable form according to the International Virtual Observatory and the WFPDB standards are discussed.

1. INTRODUCTION

There are 414 archives known in the world, which contain more than 2.1 million wide-field images (of which about 64,000 – are spectral images). The largest number of images is stored in the archives of the Harvard Observatory – more than 500 thousand. In the Sonneberg Observatory there are about 270 thousand. The total amount of plates in the Odessa collections is more than 100 thousand that covered period of 1909-1998. The key difference of the Odessa collection from others is that the images are made in two spectral bandwidth –

photographic and photovisual, that makes possible to determine not only magnitudes but also color indices of celestial objects.

2. ODESSA PLATE ARCHIVE

The Odessa archive consists of three collections: the Simeiz collection, the Old collection and the collection obtained on the Seven-camera astrograph.

2.1. Simeiz Collection (1909-1953)

The collection of near 8,000 plates was obtained at the Simeiz Observatory on Double 120-mm astrograph (Fig. 1) during the period 1909-1953. In 1966, the archive of the Simeiz observatory was moved from Crimea to Odessa by agreement between director of Simeiz observatory A.B. Severnyy and director of Odessa observatory V.P. Tsesevich for study the variable stars on these plates. Thus, the Simeiz collection became part of the Odessa archives of astrophotonegatives.

Figure 1: Double astrograph with 120 mm Unar Lenses by Carl Zeiss (worked since 1908).

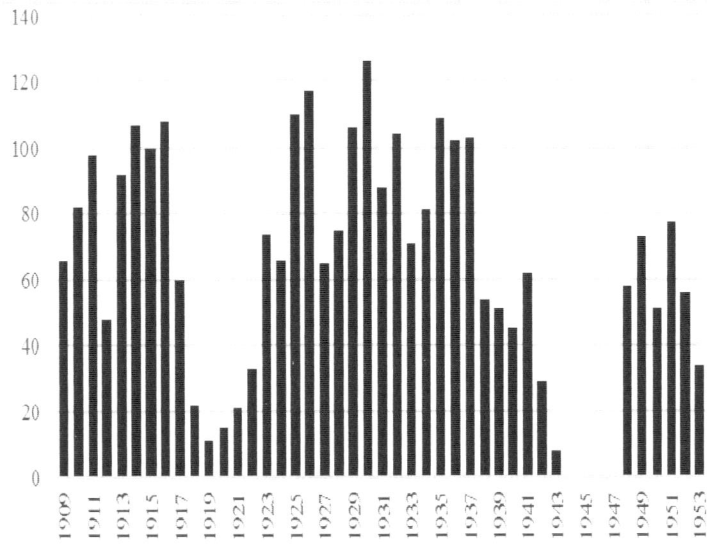

Figure 2: Annual number of observation nights in Simeiz from 1909 to 1953.

Characteristics of the Simeiz Double astrograph:
- *locations of astrograph:* Simeiz, Kitab
- *Lenses:* Two 120 mm Unar lenses by Carl Zeiss
- *operation period*: from JD 2418405 to 2434725 (1909-1953) (Fig. 2)
- *plates size*: 127 x 178 mm (5x7 inches)
- *emulsions*: more than 10 varieties of photo emulsions
- *field of view*: 11.9 x 16.2 deg
- *limiting photographical magnitude*: ~ 15^m
- *exposure time*: up to 2 hours
- *number of plates*: about 8,000

During the World War II in evacuation (1942-1944) in Kitab (Uzbekistan), observations were made on the Double astrograph by Bredikhin.

The Simeiz collection of plates contains mainly observations of small bodies of the Solar system (asteroids), Major planets and their satellites. Several hundred plates contain images of various objects and phenomena: comets, variable stars, lunar eclipses, etc. The field centers are determined by the position of small planets and comets and by their expected region of presence. Coordinates are specified on plates envelopes.

2.2. The Old Collection (1951-1957)

The Old collection of starry-sky imagery was obtained on three instruments: "Large" astrograph, "Small Two-camera" Astrograph and Three-camera astrograph "Hedgehog". During working of these three astrographs in 1951-1957, more than 10,000 plates were obtained, first of all, for studying of variable stars.

In pre-war years, the camera "Large" astrograph" with a long focus was installed on the telescope-refractor by T. Cooke & Sons (165 mm) (Fig. 3).

The cycle of patrol observations on this astrograph was organized already in the first years of his directorate in 1945, V. P. Tsesevich (1907-1983). The "Large Astrograph" was replaced by a "Small Astrograph", which already had two cameras, and then the "Three-camera" astrograph ("Hedgehog"), with short-focus cameras.

Figure 3: T. Cooke & Sons refractor with “Large” astrograph.

Characteristics of old astrographs of Odessa Observatory:

- *location of astrographs*: Odessa, Taras Shevchenko park
- *period of observations:* from JD 2433812 to 2435776 (1951-1956)
- *plates sizes*: 130x180, 180x180 and 180x240 mm
- *emulsions*: Ilford, Agfa Astro, "Isoorto" with yellow, red filters and without filters
- *field-of-view*: 24x33 degrees
- *number of guide stars*: 64 (+35 single stars)
- *limiting photographical magnitude*: ~ 13.5^m
- *exposure times*: from 0.5 to 3 hours
- *storage of plates*: observation station Mayaki

Table 1: Characteristics of astrographs "Old collection" of Odessa Observatory (Karetnikov et al. 1994).

Instrument	No. of cameras	Lenses type	D, mm	F, mm	FoV, °	Photom. System	FoV center shift from guide star, °
"Large" astrograph	1	Tessar SOI	148	1000	9 x 12	pg	0
"Small 2-camera" astrograph	1 2	Industar-17 Zeiss Triplet	100 100	500 500	22 x 15 22 x 15	pg, pr pg, pv	0 0
3-camera astrograph "Hedgehog"	1 2 3	Industar-163 Xenon Epostar	67 60 60	300 120 120	20 x 30 20 x 30 20 x 30	pv pv pv	0 -25 +25

In the "Old Collection" there is a small number of plates exposed in photored light (pr). For these plates different guide stars were chosen. These stars are specified on the plates envelopes. The plate centers listed in Table 1 are given as offset relative to the corresponding guide star (in degrees of declination).

2.3. Collection of the Seven-Camera Astrograph (1957-1998)

The largest collection in the Odessa archive contains about 84,000 photographic plates (12,225 exposures) obtained on the Seven-camera astrograph (Fig. 4) over 40 years of observations. The Seven-camera astrograph was created in 1957 during the preparation for the International Geophysical Year (IGY) at the suburban astronomical station in Mayaki (40 km from the city of Odessa) by initiative of V.P. Tsesevich. The collection was created for observing, first of all, variable stars. In addition, twilight phenomena, comets, asteroids, satellites, quasars (for example, 3C 273) were photographed, the outburst of gamma-busters and others was investigated.

Characteristics of the Seven-camera astrograph:

- *location of 7-camera astrograph*: Odessa region, Belyaevsky district, Mayaki, Observation Station of Mechnikov Odessa National University
- *observation period*: 1957-1998 years
- *plates sizes*: 130x180, 180x240 mm
- *emulsions*: Agfa Astro, ORWO ZU1, ZU-2, ZU-21without filters, ZP-1, ZP-3 with yellow filters
- *field-of-view*: 30 x 80 deg
- *guide stars*: main stars – 39, more than 75 were used from 1 to 15 times
- *limiting photographical magnitude*: ~ 14.5^m

- *limiting photovisual magnitude*: ~ 12^m
- *exposure time:* 30 min
- *number of plates*: about 84,000

Figure 4: The Seven-camera astrograph.

The astrograph was twice modernized. Some cameras changed to more efficient ones, their fields of view also changed and shifted. Thus, all the photographic plates of the Seven-camera astrograph are divided into 3 series: the Old (1957-1959), the New (1959-1966) and the Third (1966-1998) series (Table 2).

Photographic materials were produced by a German firm, which first produced photographic materials called Agfa Astro, and then - ORWO. There were only two types of photographic materials: non-sensiblised ZU-1, ZU-2 and ZU-21 and panchromatic ZP-1 and ZP-3, which gave homogeneous photometric systems. All changes of photometric systems were investigated in the work of A. N. Rudenko (Rudenko, 1988).

Table 2: Characteristics of the Seven-camera astrograph.

Name of series and dates	No. of camera	Type of objective	D, mm	F, mm	Field-of-view, °	Photom. system	FoV center shift from guide star, °
The Old series. From 19.07.1957 to 08.04.1959	1	Uran-9	100	250	30 x 40	pg, pv	+13
	2	Uran-9	100	250	30 x 40	pg, pv	-13
	3	Uran-9	100	250	35 x 25	pg	+25
	4	Vierlinser	160	720	13 x 18	pg	+08
	5	Vierlinser	160	720	13 x 18	pg	-08
	6	Uran-9	100	250	30 x 40	pg	-27
	7	Tessar SOI	148	100	9 x 12	pg	0
The New series. From 09.04.1959 to 08.06.1966	1	Uran-9	100	250	30 x 40	pv	+10
	2	Uran-9	100	250	30 x 40	pv	-10
	3	Triplet	100	500	22 x 15	pg	+30
	4	Vierlinser	160	720	13 x 18	pg	+18
	5	Vierlinser	160	720	13 x 18	pg	-18
	6	Industar 17	100	500	22 x 15	pg	-33
	7	Uran-12	200	500	18 x 24	pg, pv	0
The Third series. From 09.06.1966 to 31.10.1998	1	Uran-9	100	250	30 x 40	pv	+13
	2	Uran-9	100	250	30 x 40	pv	-13
	3	UNAR	120	600	16 x 11	pg	+23
	4	Vierlinser	160	720	13 x 18	pg	+08
	5	Vierlinser	160	720	13 x 18	pg	-08
	6	UNAR	120	600	16 x 11	pg	-23
	7	Uran-12	200	500	18 x 24	pv	0

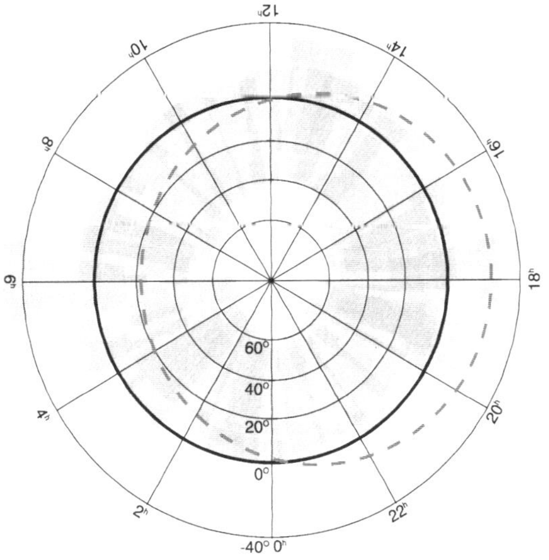

Figure 5a: Celestial sphere covering by fields of view of cameras No. 3, 4, 5 and 6, The Third series.

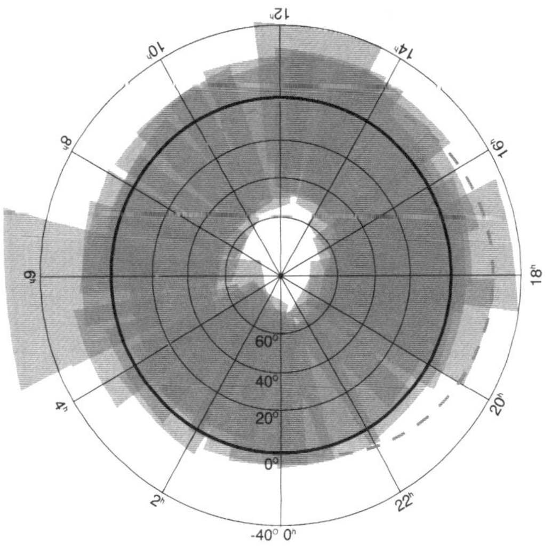

Figure 5b: Celestial sphere covering by fields of view of cameras No. 1, 2 and 7, The Third series.

The yellow filters were installed on the 1st and 2nd cameras from July 12, 1958 and on the 7th camera from June 18, 1965.

The number of clean nights based on the data of the 7-camera astrograph (Kashuba et al., 2015) can be traced from the statistics of patrol images (Figure 6):

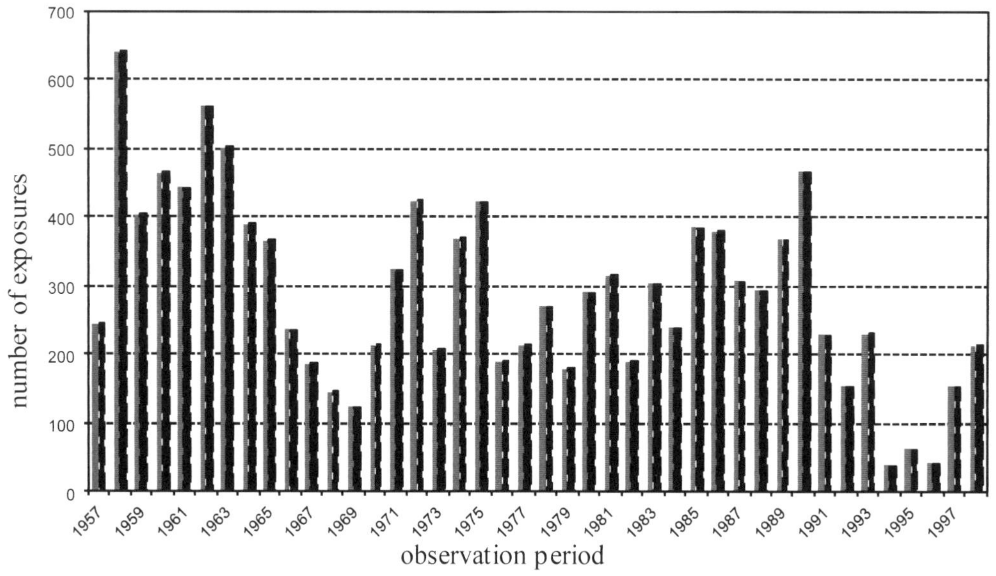

Figure 6: Statistics patrol images on the Seven-camera astrograph for the period 1957-1998.

3. DIGITIZATION OF THE ODESSA ASTROPHOTONEGATIVE ARCHIVE

The main goal of the digitization of the entire archive is to preserve the historical heritage of the Odessa Observatory, to update the archive as a resource for obtaining full photometry and astrometry results for the greater part of the northern sky, as an extension of the observatory's resource base. In other words, make the archive available to scientific community.

For the first time, the digitization of the Odessa archive of plates was started in 1999. The project of digitization was proposed by A. Yushchenko, et al. (Karetnikov et al., 2007), but was not launched for financial difficulties. In 2002, only the original metadata was prepared in the form of a machine-readable catalog recorded for 2 collections: the Simeiz and the Seven-camera astrograph (Pikhun et al, 2002).

Digitization of the Odessa archive began in 2013 in the Main Astronomical Observatory of National Academy of Science of Ukraine using the flatbed scanner Epson Expression 10000XL. About 500 digitized 16-bit grayscale images from the Simeiz collection were obtained with resolution of 1200 dpi and included in the combined digital archive of Ukrainian Virtual Observatory Joint Observation Data Archive (JDA) (Vavilova, 2012). To date, the JDA contains 600 metadata records and more than 400 digitized images of selected plates collection in the 1950s (http://gua.db.ukr-vo.org/createlist.php?aid=CRI012A).

In the summer of 2014 to glass library of the Odessa observatory located in the village Mayaki, sponsors have presented the Epson Perfection V700 Photo scanner (the A4 format, optical resolution 6400x9600 dpi, depth of color, bit: 48) complete with office equipment serving the scanner. From this date digitization of the Odessa archive began to be conducted locally, not taking out of its far from its storage place.

The attestation of the Epson Perfection V700 Photo scanner was carried out by V. Andruk (Andruk et al, 2015). As a result, he came to the conclusion that for rectangular coordinates mean-square error of value of one difference determination is σ_{xy} = 0.023 pixels for the 1200 dpi scanning mode; for instrumental stellar magnitudes, the mean-square error value of one difference determination is $\sigma_m = 0.013^m$ (Kashuba et al, 2017).

On Epson Perfection V700 Photo scanner, up to 2018, about 4,000 plates of the Simeiz collection were digitized. In 2016 about 400 of the collection of the Seven-camera astrograph (4 cameras, γ Cyg, III series) were digitized for the purpose of researching stars for their variability in brightness. The plates were digitized in TIFF format, 16 bit, 20 microns/pixel, the file size is ~ 90 MB (for 1200 dpi resolution) and ~ 350 MB (for 2400 dpi resolution).

A parallel process for placing material in an open access database is the digitization of the corresponding observational logbooks and envelopes of plates, as well as the creation of machine-readable data for each plate in a single WFPDB format.

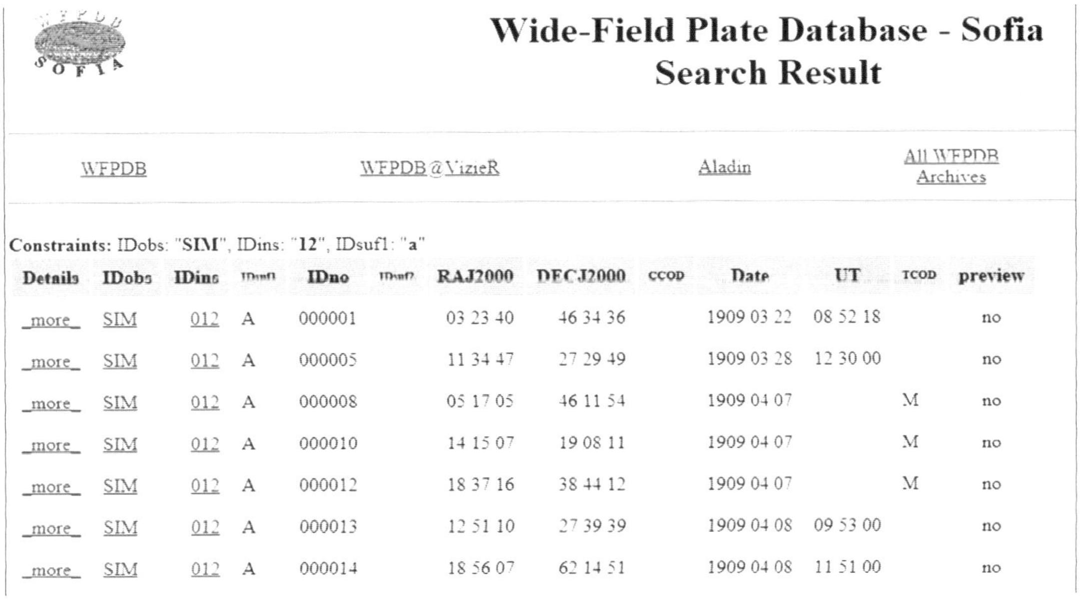
Wide-Field Plate Database - Sofia
Search Result

WFPDB | WFPDB@VizieR | Aladin | All WFPDB Archives

Constraints: IDobs: "SIM", IDins: "12", IDsufl: "a"

Details	IDobs	IDins	IDsuf1	IDno	IDsuf2	RAJ2000	DECJ2000	CCOD	Date	UT	TCOD	preview
more	SIM	012	A	000001		03 23 40	46 34 36		1909 03 22	08 52 18		no
more	SIM	012	A	000005		11 34 47	27 29 49		1909 03 28	12 30 00		no
more	SIM	012	A	000008		05 17 05	46 11 54		1909 04 07		M	no
more	SIM	012	A	000010		14 15 07	19 08 11		1909 04 07		M	no
more	SIM	012	A	000012		18 37 16	38 44 12		1909 04 07		M	no
more	SIM	012	A	000013		12 51 10	27 39 39		1909 04 08	09 53 00		no
more	SIM	012	A	000014		18 56 07	62 14 51		1909 04 08	11 51 00		no

Figure 7: Screenshot of the SIM012A catalog page in WFPDB (www.wfpdb.org).

The processing of archival data with unified standards gives objective unique information about the astronomical object. Tsvetkov et al. (Tsvetkov, 2005), using the example of the Pleiades, showed how to create an almost continuous database, overlapping in different archives, to determine the sky area from 1872 to 1996. The data obtained from the Odessa archive will be a reasonable addition to these combined structures.

In May 2018, 887 records of the Simeiz collection SIM012A (80 GB) of the period 1909-1915 were placed in the database of the Bulgarian Data Processing Center (WFPDB) (Figure 7) (Tsvetkov, 2006).

4. POSSIBLE SCIENTIFIC OBJECTIVES

Speaking about the relevance of the archive, usually given examples of scientific problems based on archival data. Let us give an example of our own photometric studies.

One of the first objects of research by processing digitized images was the star KIC 8462852, also known in the literature as the star of Tabbi.

Interest in this star is caused by irregular blackouts on the light curve discovered by Kepler's mission, as well as by unusual interpretations of such changes. Two years ago, Bradley Shafer published the results of a photometric study of the star KIC 8462852 on the plates of the Harvard archive for a period of 90 years (Schaefer, 2016). He found tendency of magnitude reducind by 0.16 mag in a hundred years (Figure 8, graph above).

A similar work was done on the plates of the Seven-camera astrograph. The lower graph (Fig. 8) shows magnitude estimates by processing 380 digitized plates of the Seven-camera astrograph. A photometric series of the star KIC 8462852 was obtained, covering interval from 1966 to 1998 - the trend of decreasing the brightness of the star is noticeable.

And in conclusion quote from page Digital Access to the Sky Century @ Harvard (DASCH): “The stars photographed on glass plates between 1880s-1990s operate as time capsules, allowing astronomers to study how the sky has changed over one hundred years”.

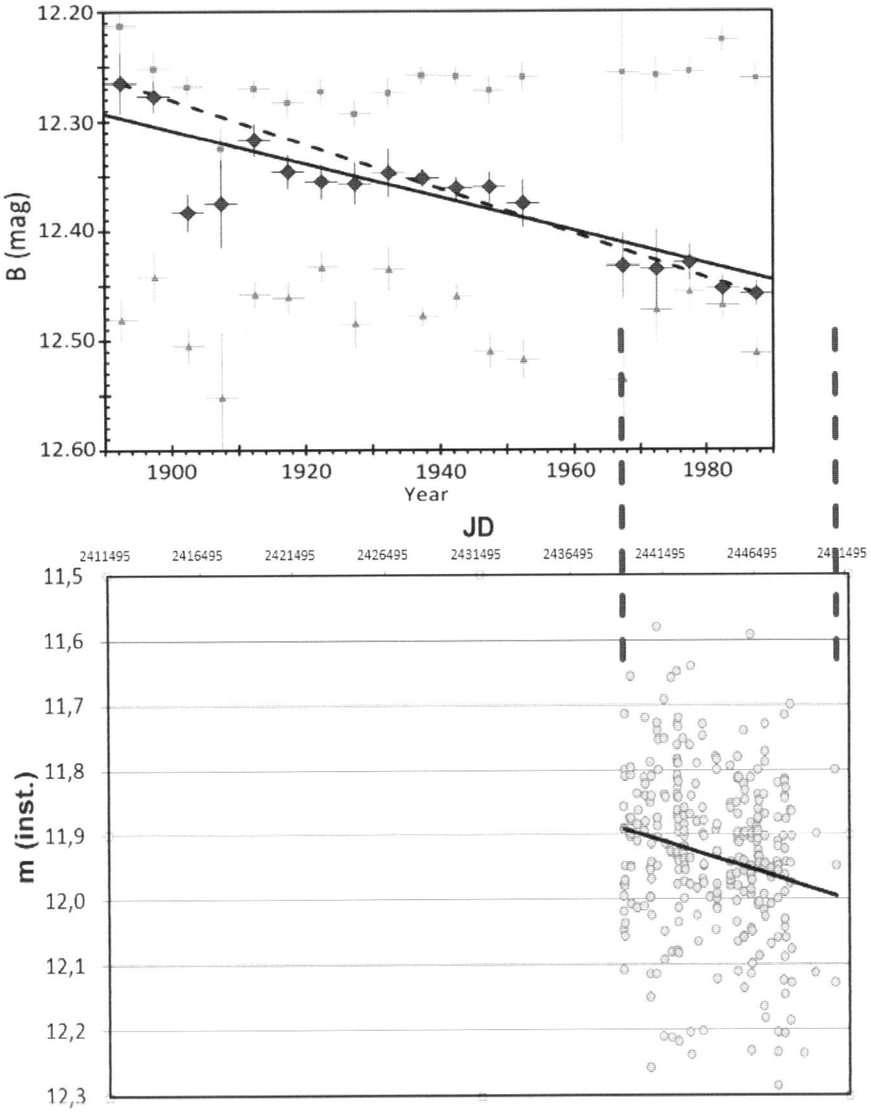

Figure 8: Photometric row on the plates of the Harvard archive (above) (https://arxiv.org/abs/1601.03256), on the plates of the seven-camera astrograph (below).

Acknowledgments

The team of authors expresses its gratitude to the organizers of the XI Bulgarian-Serbian Astronomical Conference and the Local Organizing Committee for assistance with attending the meeting, good contacts and hospitality.

References

Andruk, V. M., Pakuliak, L. K., Golovnia, V. V., et al.: 2015, *Odessa Astronomical Publications*, **28**, 192.

Karetnikov, V. G., Markina, A. K., Sotnikov, V. P.: 1994, "The Odessa Sky Patrol Plate Collection", *Flares and Flashes*; Proceedings of the IAU Colloquium No. **151**; Held in Sonneberg; Germany; 5 - 9 December 1994; XXII; 477 pp., Springer-Verlag Berlin Heidelberg New York. Also *Lecture Notes in Physics,* **454** Edited by Jochen Greiner, Hilmar W. Duerbeck, and Roald E. Gershberg, p.407.

Karetnikov, V. G., Dorokhova, T. N., Yushchenko, A. V., Pikhun, A. I., Koshkin, N. I.: 2007, "The Digitization Project of Odessa Plate Depository", Library and Information Services in Astronomy V, A*SP Conference Series*, **377.**

Kashuba, S., Gorbanev, Yu., Andrievsky, S., Marsakova, V., Troianskyi, V.: 2015, "The quantity and quality of observational nights monitored with using the astronomical instruments at the suburban observation stations of astronomical observatory of Odessa National University", *Odessa Astronomical Publications*, vol. **28/1**, 16.

Kashuba, S., Andruk, V., Kashuba, V.: 2017, *Odessa Astronomical Publications*, **30**, 174.

Pikhun, A. I., Yushchenko, A. V.: 2002, *IBVS,* 5215.

Rudenko, A. N.: 1988, *Kinematics and Physics of Celestial Bodies*, **4**, 34.

Schaefer, B.: 2016, „KIC8462852 Faded at an Average Rate of 0.165±0.013 Magnitudes Per Century From 1890 To 1989", arXiv:1601.03256v1.

Tsvetkov, M., Tsvetkova, K., Borisova, A.: 2005, *Kinematics and Physics of Celestial Bodies*, Suppl., **5**, 567.

Tsvetkov, M.: 2006, "Wide-Field Plate Database: a Decade of Development", *Virtual Observatory: Plate Content Digitization, Archive Mining and Image Sequence Processing, iAstro workshop*, Sofia, Bulgaria, 2005, ISBN-10 954-580-190-5, p. 10.

Vavilova, I. B., Pakulyak, L. K., Shlyapnikov, A. A., Protsyuk, Yu. I., Savanevich, V. E., Andronov, I. L., Andruk, V. N., Kondrashova, N. N., Baklanov, A. V., Golovin, A. V., Fedorov, P. N., Akhmetov, V. S., Isak, I. I., Mazhaev, A. E., Golovnya, V. V., Virun, N. V., Zolotukhina, A. V., Kazantseva, L. V., Virnina, N. A., Breus, V. V., Kashuba, S. G., Chinarova, L. L., Kudashkina, L. S., Epishev, V. P.: 2012, Astroinformation resource of the Ukrainian virtual observatory: Joint observational data archive, scientific tasks, and software. *Kinematics and Physics of Celestial Bodies*, **28(2)**, 85-102 (04/2012).

Proceedings of the XI Bulgarian-Serbian Astronomical Conference
(XI BSAC) Belogradchik, Bulgaria, May 14-18, 2018
Editors: Milcho K. Tsvetkov, Milan S. Dimitrijević and Momchil Dechev
Publ. Astron. Soc. "Rudjer Bošković" No 18, 2018, 87-92

NEW CAPABILITIES OF THE SOFTWARE TO SUPPORT DIGITIZATION OF ASTRONOMICAL PHOTOGRAPHIC PLATES

NIKOLAY KIROV[1] and MILCHO TSVETKOV[2,3]

[1]*Computer Science Department, New Bulgarian University*
Montevideo Str. 21, Sofia 1618, Bulgaria
[2]*Astronomical Institute with National Astronomical Observatory, Bulgarian Academy of Sciences*
72, Tsarigradsko Chaussee Blvd., Sofia 1784, Bulgaria
[3]*Institute of Mathematics and Informatics, Bulgarian Academy of Sciences*
Acad. Georgi Bonchev Str., Block 8, Sofia 1113, Bulgaria
E-mail: nkirov@nbu.bg, tsvetkov@skyarchive.org

Abstract. We In this article, we announce improvements to two applications for digitizing astronomical photographic plates and developing a new one. In the software for creating the header of fits file using the data from WFPDB (wfpdb.org) we add feature for processing data for multiexposure plates. A new functionality of application for updating fits header allow us to complete fits header for fits files obtained from ImageMagic application. The development of virtual observatories has prompted us to realise a convertor of fits header to GAVO (www.g-vo.org) standard.

1. WIDE-FIELD PLATE DATABASE

There are about 2475000 astronomical photographic plates all over the world. Wide-Field Plate Database (www.wfpdb.org) is WEB-based database that contains meta-data for more than 600 thousand plates (see Kalaglarsky et al: 2014). It is one wide-field unique telescope, giving access to unique photographic astronomical observations, done systematically in the period 1880 – 2000!

The Catalogue of Wide-Field Plate Indexes contains meta-data for plates. For every plate many details are stored in the database: the coordinates of the plate centre, the date and time of the observation, object name and type, method of observation, duration of exposures, type of emulsion, the size of the plate, the quality of the plate, the name of the observer, etc.

The meta-data of the plates are distributed in 6 plain-text files. For every file we give an example – one line in a file contains the data for a plate with identifier ESO040 007863.

- The most important information is in the maindata file:
ESO040 007863 053517-052328 19850114012100 ORION M42-43 NGC1976 S42 6120.0KODAK IIaO Pg161611111
- Availability file:
ESO040 007863 PLATE IS AVAILABLE AT THE WFPDB DEPOSITORY, SOFIA, BULGARIA: milcho.tsvetkov@gmail.com

- Notes file:
ESO040 007863 ORIG_COORD:053400-052400 ST.ST= ST.END=0620 UT.ST=0121 UT.END=0329 MULTIP. EXPOS=6X20 FL! ANON L EXP1-6_UT_STARTUT_END=01:21-01:41,01:42-02:02,02:03-02:23,02:27-02:47,02:48-03:08,03:09-03:29;

- Observer file:
ESO040 007863 W.SEITTER

- Quality file:
ESO040 007863 3 ARCSEC SEEINGS

- Digitization file:
ESO040 007863 DIGITISED WITH PDS2020 AT AIM, AND IN THE WFPDB WITH THE FB SCANNER EPSON PERFECTION V700 20MIC PIXEL(1200DPI)

2. PLATE DIGITIZATION

Plate image is a digitized astronomical photographic plate image (16-bit grayscale FITS file), with high resolution aiming photometric and astrometric measurements.

The process of plate digitization consists of 3 steps:
• Scanning the plate (image) to produce .tif (row-tiff) file (see Tsvetkov et al: 2012).
– Hardware: flatbed scanner and – Software: VueScan
• Completing metadata (header) to produce .hdr file
– Transferring data from WFPDB: header application and
– Adding actual data for scanning, time, etc.: header application
• Completing header file and merging it with image file to produce .fits file
– Tif2fits application or
– ImageMagic software + UpdateFitsHdr application

3. HEADER

The FITS header contains the complete metadata for the plate, the telescope, the place of the observation, the scanning process, etc. The main part of these data can be extracted from the corresponding WFPDB catalogues. This approach presumes that the collection of plates for digitization (plate archive) is already described in WFPDB. The second part of the data consists of constant values, and the third part can be calculated using the other parts.

.hdr denotes a plain text file producing by FITS header application.

• Most of the data for fits header are extracted from WFPDB.

• Additional data are entered automatically using config file for the processing archive

• All the data can be corrected manually.

Figure 1 presents the general view of FITS header application.

The format of the data in the FITS header is strictly defined (see FITS header format: 2014 and FITS Standard Document: 2016) but the extended content is very flexible. There are some specific data for photographic plates, comparing with CCD images. Our choice of key words, fields and measurement units is defined from practice with various plate archives (see Kirov et al: 2012b). The development of virtual observatories (VO) during last years has prompted us to adapt out software for FITS headers for facilitation the publication of digitized archives in VO.

The important place for our digitized archives is German Astrophysical Virtual Observatory (GAVO) and our goal is to satisfy the requirements of GAVO.

GAVO header sections are:

- "Basic data"
- Original data of observation • Observatory and instrument • Photographic plate
- Derived observation data
- Scan details
- Data files

The data from WFPDB is transferred to .hdr file but the complete FITS header contains also information for image and we denote this file as .hdrf file (for easy on-line access to the plate meta- data.). If we denote by .hdrg GAVO header, we need the following convertors for various "standards":

- WFPDB → .hdr • .hdr → .hdrg
- .hdrf → .hdrg

Example of differences between two formats:

.hdrf: UT = '1911-02-02 01:31:39' / date and UT at mean epoch

.hdrg: DATE-AVG= '1911-02-02T01:31:39Z' / UT d/t mid-point of observation

4. PROCESSING IMAGE

While VueScan produces .tif image format files, the conversion to .fits image format (including header) can be done in two ways – using Tif2fits application or using a combination of ImageMagic software and our UpdateFitsHdr application. Tif2fits transforms .tif and .hdr files to .fits and .hdrf files (Figure 2).

ImageMagic converts .tif to .fits file with very small FITS header. UpdateFitsHdr application replaces this small header using .hdr file and produces normal .fits file and .hdrf header file (Figure 3).

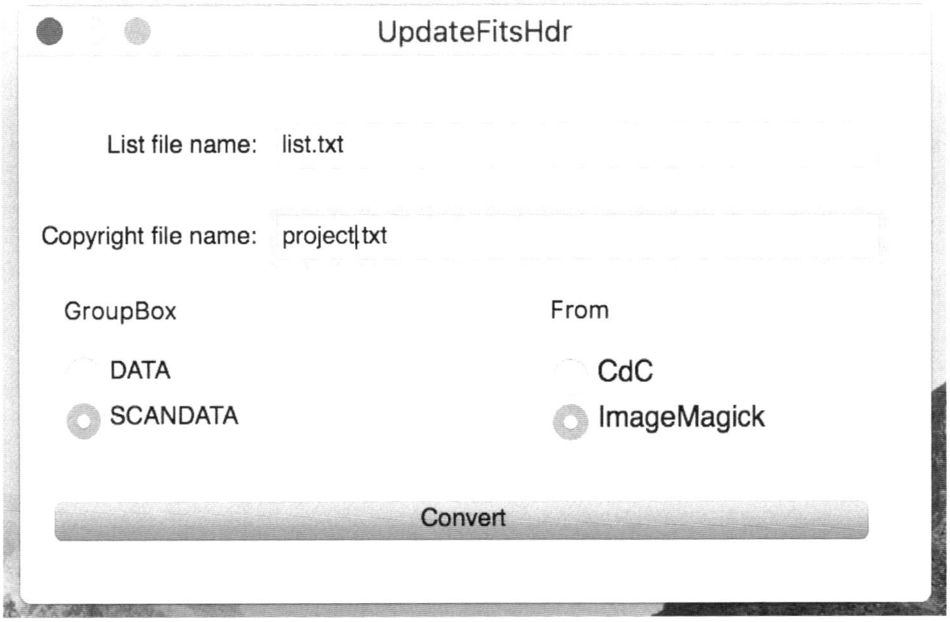

5. MULTIPLE EXPOSURE

Multi exposure plates contain several exposures and its header has to contain data for every single exposure. This way the header turns to variable number of fields (lines), which does not match to FITS header application. MultiExp application solves the problem extending .hdr or .hdrf file (Figure 4).

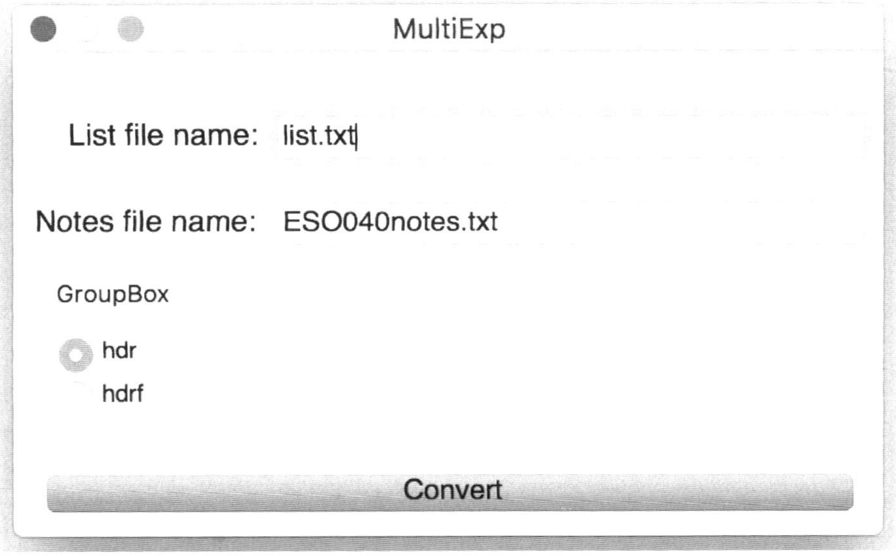

The data for every exposure can be found in WFPDB – Notes file.
Example of a part of a line for one plate in WFPDB and the same data in FITS header for that plate.

WFPDB Notes file:
UT_STARTUT_END=01:21-01:41,01:42-02:02,02:03-02:23,02:27-02:47,02:48-03:08,03:09-03:29;

.hdrg file:
DT-OBS1 = '1985-01-10T01:21:00Z'
DT-AVG1 = '1985-01-10T01:31:00Z'
DT-END1 = '1985-01-10T01:41:00Z'
EXPTIM1 = 1200.0
DT-OBS2 = '1985-01-10T01:42:00Z'
...

6. CONCLUSIONS

The software is written in C++ using Qt – cross-platform application and UI development framework (www.qt.io). The applications are publicly available in GitHub as open source (github.com/nkirov).

References

Kalaglarsky, D., Tsvetkova, K., Tsvetkov, M.: 2014, „WFPDB: Import of Catalogue of Wide-Field Plate Archives v7.0 and Recent Development", Proceedings of the IX Serbian-Bulgarian Astronomical Conference (IX SBAC), Sofia, Bulgaria, July 2-4, 2014, Editors: M. S. Dimitrijević and M. K. Tsvetkov, *Publ. Astron. Soc. "Rudjer Bošković"*, **15**, 55-62.

Kirov, N., Tsvetkov, M., Tsvetkova, K.: 2012a, „Software Tools for Digitization of Astronomical Photographic Plates", *Serdica Journal of Computing*, **6 (1)**, 67-76.

Kirov, N., Tsvetkov, M., Tsvetkova, K.: 2012b, „Technology for digitization of astronomical photographic plates, Proceedings of the 8th Annual International Conference on Computer Science and Education in Computer Science, Boston, USA, 5-8 July 2012, Publ. New Bulgarian Uiniversity, 109-114.

Tsvetkov, M., Tsvetkova, K., Kirov, N.: 2012, „Technology for scanning of astronomical photographic plates", *Serdica Journal of Computing*, **6 (1)**, 77-88.

FITS header format: 2014, https://www.plate-archive.org/wiki/index.php/FITS_header_format

FITS Standard Document: 2016 https://fits.gsfc.nasa.gov/fits_standard.html

Proceedings of the XI Bulgarian-Serbian Astronomical Conference
(XI BSAC) Belogradchik, Bulgaria, May 14-18, 2018
Editors: Milcho K. Tsvetkov, Milan S. Dimitrijević and Momchil Dechev
Publ. Astron. Soc. "Rudjer Bošković" No 18, 2018, 93-98

MAGNETIC FIELD VARIABILITY IN RZ ARI – A FAIRLY EVOLVED M GIANT

RENADA KONSTANTINOVA-ANTOVA[1], AGNES LEBRE[2], MICHEL AURIERE[3], RUMEN BOGDANOVSKI[1], SVETLA TSVETKOVA[1], ANA BORISOVA[1], PHILIPPE MATHIAS[3], BENDJAMIN THESSORE[2], RADOSLAV ZAMANOV[1] and SVETLANA BOEVA[1]

[1]*Institute of Astronomy and NAO, Bulgarian Academy of Sciences*
72 Tsarigradsko shosse blvd., Sofia 1784, Bulgaria
[2]*LUPM, University of Montpellier, CNRS, Montpellier, France*
[3]*IRAP, 14 avenue Édouard Belin, 31400, Toulouse, France*
E-mail: renada@astro.bas.bg

Abstract. RZ Ari is a fast rotating apparently single M giant of 2.2 Msun. It is fairly evolved to tip RGB or early AGB stage. In addition, the star is known as semi-regular variable. We have studied its longitudinal magnetic field variability using spectropolarimetric data obtained with Narval at Telescope Bernard Lyot, Pic du Midi Observatory, France in the period 2008 – 2018. Two periods were identified using the Lomb – Scargle method: 1310 days and 498 days. The second one is very close to the Long Secondary Period of RZ Ari (480 days) and maybe we observe for first time an interplay of the magnetic field and pulsations for a M giant. The first period is rather possible due to rotation modulation caused by the surface magnetic structures. Taking into account literature data we determined the radius of the star (~117 Rsun) that is consistent with the AGB phase. On the basis of the observed period of 1310 days and the calculated Psini (using radius of 117 Rsun and Vsini = 9.6 km/s) we found an inclination angle for the star of about 29 degrees.

Our work hypothesis is that RZ Ari with its fast rotation is an intermediate case of dynamo generated magnetic field and shock wave compression generated fields as we observe in Miras. These stars are the next evolutionary stage, after early-AGB stars. Further study is required to confirm or reject the hypothesis and to understand better the interplay of the magnetic field and pulsation in this fairly evolved giant.

1. INTRODUCTION

Recently, magnetic fields (MF) were detected in many single G, K and M giants (Konstantinova-Antova et al. 2013, 2014; Aurière et al. 2015). While the reasons for the MFs generation in G,K giants is mostly α-ω dynamo and remnant MF in the Ap star descendants, for the M giants the reasons are not completely clear. Charbonnel et al. (2017) consider that α-ω dynamo could operate even in early assymptotic giant branch stars (AGB) due to the properties of their convective envelopes. In addition, some of these stars possess faster rotation that could not be explained by the theory of the stellar evolution, yet. On the other hand, most of these M giants are semi-regular variable stars with pulsations and it is still unclear how these pulsations relate to the MF generation. What is also found recently for the magnetic M giants is that they occupy a certain area on the Hertzsprung-Russel diagram (HRD), the so-called "second magnetic strip" (Konstantinova-Antova et al. 2014). This strip coinsides with the tip of the red giant branch (RGB) and early-AGB phase, in agreement with the theoretical models of Charbonnel et al. (2017).

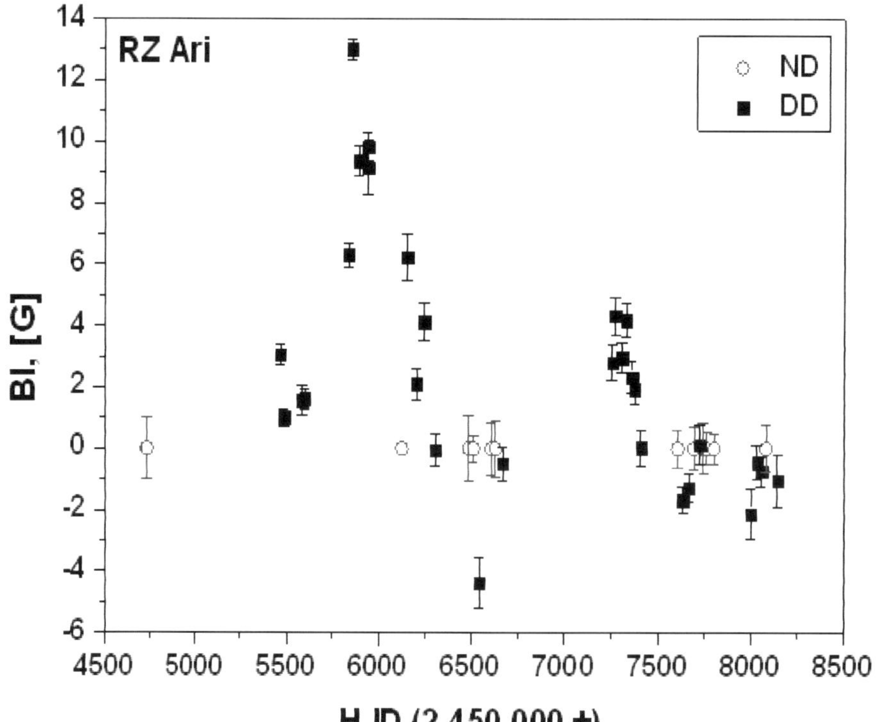

Figure 1: Bl variability of RZ Ari in the period Sept. 2008 – Jan. 2018. The black filled squares stand for magnetic field definite detection (DD) and the red open circles stand for non-detection (ND).

RZ Arietis = HD 18191 is a 6 mag single star of M6 III spectral class. It is the star with largest Vsini = 9.6 km/s and the strongest longitudinal MF (Bl = 13 G) among all M giants studied by our group. According to Konstantinova-Antova et al. (2010) its effective temperature Teff = 3450 K and log (L/Lsun) = 3.11. The star is of about 2.2 Msun and is situated either on the tip RGB or in the beginning of AGB. On the other hand, taking into account the measured angular diameter for RZ Ari of 0".01022 (Richichi et al. 2005) and the distance to the star of 107.76 pc (van Leeuven et al. 2007) we obtain a radius of 117.2 Rsun that is consistent with the AGB phase.

Also, RZ Ari is known as a semi-regular variable star with a period of pulsations of about 50 days and a Long Secondary Period (LSP) of 480 days (Percy et al. 2008, 2016; Tabur et al. 2009). Here we present the first results of about 10 years of spectropolarimetric observations and magnetic field study for this fairly evolved giant with fast rotation.

2. OBSERVATIONS AND DATA PROCESSING

The observations were carried out with the 2m Bernard Lyot telescope (TBL) at Pic du Midi Observatory, France. The telescope is equipped with Narval spectropolarimeter (Aurière, 2003). Narval has a spectral resolution in polarimetric mode of 65 000 and covers the spectral region of 360 to 1000 nm. Stokes I (unpolarized) and Stokes V (circular polarization) parameters were obtained. The technique of observations is similar to the one described in detail in Konstantinova-Antova et al. (2010). For the Zeeman analysis of the observations we applied the Least-Square Deconvolution technique (LSD, Donati et al. 1997). We used a mask calculated for solar chemical abundances and consistent with the effective temperature and log g of RZ Ari. The LSD method enables us in the case of RZ Ari to average more than 12 000 spectral lines to get the Stokes I and V profiles. The so-called "null spectrum" (obtained as a result of rotation of the retarders in the instrument, the Freshnel rhombs) given by the standard procedure was also examined, but no signal was found. That confirms that the detected signatures in the V profile are not spurious ones. From the obtained mean LSD profiles for each night the surface-average longitudinal MF Bl was computed (Rees & Semel 1979; Donati et al. 1997). For RZ Ari the typical Bl accuracy is of 0.5 G.

3. RESULTS AND DISCUSSION

A strong MF variability was found for RZ Ari in the period 2008-2018 (Fig.1). The longitudinal MF varied in the interval +13 G to -5 G with short periods of non-detection. This MF behavior is typical for the magnetic M giants (Konstantinova-Antova et al. 2013) and is different to what we observe in the magnetic G and K giants (Konstantinova-Antova et al. 2013; Aurière et al. 2015). A possible explanation is a non-uniform surface distribution of the magnetic areas. Lomb – Scargle method (Scargle, 1982) was applied for period search to the whole Bl dataset. A period of 1310 +85/-73 days was determined with a false alarm probability (fap) of 0.3%. In addition, a second, shorter period of 498 +8/-5 days was identified with a fap of 35% (Fig. 2). The phased Bl regarding the period 1310 days is presented in Figure 3. The shorter period is close to the LSP (~480 d) determined on the basis of photometric observations for RZ Ari (Percy et al. 2008, 2016). Such LSP is typical for the semi-regular variable stars, but its origin remains still unclear. Percy et al. (2016) give some assumptions for it like turnover of giant convection cells; oscillatory convective modes; dusty cloud orbiting the red giant; rotation modulation due to spots. Also, no more than one LSP is known for the stars they have studied. What could be concluded by the analysis of the published information: 1. in many cases Prot > LSP, hence rotation modulation is not likely; 2. the LSP color variations from photometry are similar to those of the pulsation period and with a sine shape (in the case of RZ Ari P ~ 50 d), and in this way the assumption for a dusty cloud and rotation modulation are excluded. In addition, contrary to the supergiant Betelgeuse (Aurière et al. 2016; Mathias et al. 2018) no evidences for giant convection cells are found for early-AGB stars. We have no idea if the convective oscillatory modes produce color variations similar to pulsations. It seems more natural that the LSP origin relates to pulsations and in this case we observe the interaction between the MF and pulsations of some kind in RZ Ari. However, more observations and their analysis are required to confirm this assumption.

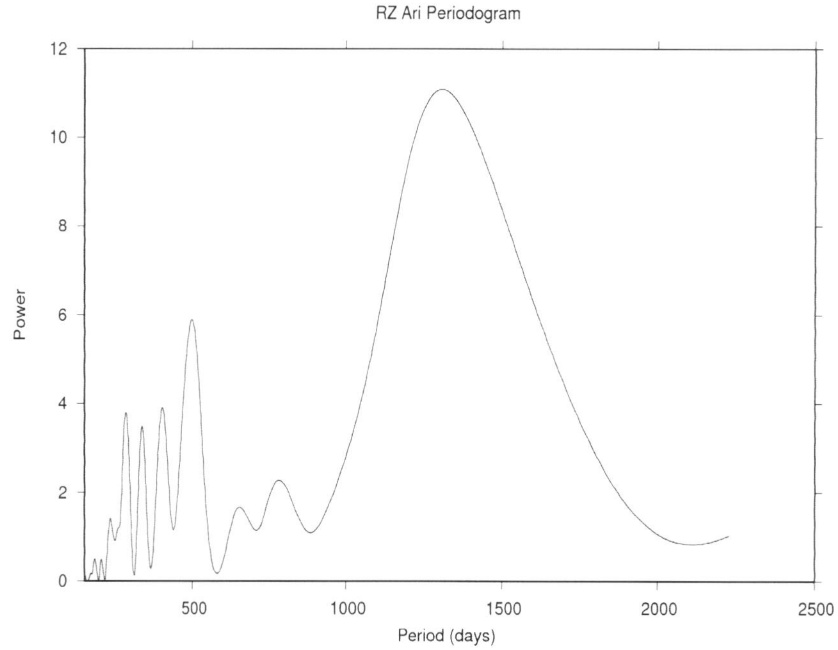

Figure 2: RZ Ari periodogram applying the Lomb-Scargle method. Two periods are identified – 1310 days and 498 days. The fap for the first one is 0.3% and for the second one it is 35%.

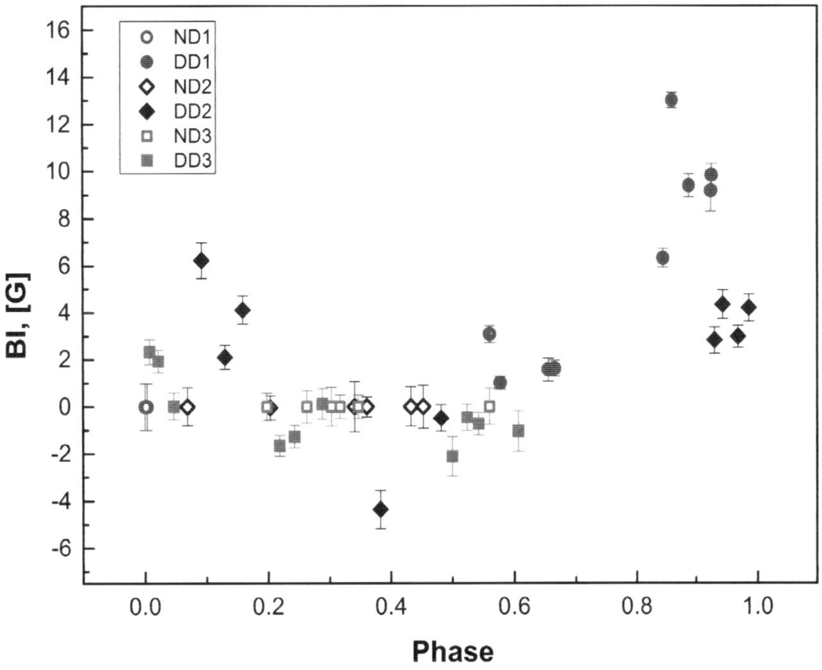

Figure 3: Bl variability of RZ Ari phased regarding the period of 1310 days.

For the longer Bl periodic variability the question is whether it is due to rotation modulation. Taking into account the Vsini and radius values mentioned in Sect.1 and the determined period, in the case of rotation modulation we obtain inclination angle i ~ 30°. That means, the star is rather tilted and in such a case the south pole is not observed. This assumption is consistent with the strong positive polarity we observe and the weaker negative one. Our next aim is to perform Zeeman Doppler Imaging (ZDI) for this M giant.

4. CONCLUSIONS

The longitudinal magnetic field variability of RZ Ari was studied on the basis of spectropolarimetric observations obtained in the period Sept. 2008 – Jan. 2018.

Two periods were identified: a long period of 1310 days and a shorter one of 498 days. The last one is very close to the LSP determined on the basis of photometric observations. After analysis of the eventual hypotheses for the origin of the LSP our conclusion is that for RZ Ari it is rather related to some kind of pulsations. If this is the case, maybe we observe for the first time an interplay between the MF and pulsations for a M giant. The longer period of 1310 days we determined is rather possible due to the rotation modulation and the inclination angle of the star (determined by using the calculated Psini) is about 29 degrees.

Taking into account that RZ Ari is an early-AGB star, but fast rotating, our work hypothesis is that it appears a case of transition between pure dynamo generated MFs and the shock wave compressed MFs found in the AGB Mira star χ

Cyg that is just at the next evolutionary stage after RZ Ari (Lèbre et al. 2014). A further study is required to verify this assumption.

Acknowledgements

We thank the TBL team for the service observing. The observations are granted with observational time under two OPTICON projects for semesters 2008B and 2011B, the Bulgarian NSF project DSAB 01/3 in 2010, the EU project BG 051PO001-3.3.06-0047 in 2012 and 2013, and the French PNPS program for the period 2014-2018. R.K.-A., R.B. and S.Ts. acknowledge partial financial support under Bulgarian NSF contract DN 18/2.

References

Aurière, M. 2003: in "*Magnetism and Activity of the Sun and Stars*", Eds J. Arnaud and N. Meunier, EAS Publ. Series **9,** 105.
Aurière, M., Konstantinova-Antova, R., Charbonnel, C. et al. : 2015, *A&A*, **574**, 90.
Aurière, M., Lopez Ariste, A., Mathias, P. et al.: 2016, *A&A*, **591**, 119.
Charbonnel C., DecressinT., Lagarde N. et al.: 2017, *A&A*, **605**, 102.
Donati, J.-F., Semel, M., Carter, B. D., Rees, D. E., Collier Cameron, A.: 1997, *MNRAS*, **291**, 658.
Konstantinova-Antova, R., Aurière, M., Charbonnel, C. et al.: 2010, *A&A*, **524**, 57.
Konstantinova-Antova, R., Aurière, M., Charbonnel, C. et al.: 2013, *Bulgarian Astronomical Journal*, **19**, 14.
Konstantinova-Antova, R., Aurière, M., Charbonnel, C. et al.: 2014, in *Proc. IAU Symposium* **302,** 373.
Lèbre, A.; Aurière, M.; Fabas, N. et al.: 2014, *A&A*, **561**, 85.
Mathias, P., Aurière, M., Ariste, A., López, et al.: 2018, *A&A*, **615**, 116.
Percy, John R., Mashintsova, M., Nasui, C. O. et al.: 2008, *PASP*, **120**, 523.
Percy, J. R., Deibert, E.: 2016, *JAVSO*, **44**, 94.
Rees, D. E., Semel, M. D.: 1997, *A&A*, **74**, 1.
Richichi, A., Percheron, I., Khristoforova, M.: 2005, *A&A*, **431**, 773.
Scargle, J.: 1982, *ApJ,* **263**, 835.
Tabur, V., Bedding, T. R., Kiss, L. L. et al.: 2009, *MNRAS*, **400**, 1945.
van Leeuwen, F.: 2007, A&A, **474**, 653.

Proceedings of the XI Bulgarian-Serbian Astronomical Conference
(XI BSAC) Belogradchik, Bulgaria, May 14-18, 2018
Editors: Milcho K. Tsvetkov, Milan S. Dimitrijević and Momchil Dechev
Publ. Astron. Soc. "Rudjer Bošković" No 18, 2018, 99-107

FIRST SERBIAN WORKS ON THEORY OF RELATIVITY

ŽARKO MIJAJLOVIĆ, NADEŽDA PEJOVIĆ
and VIKTOR RADOVIĆ

Faculty of Mathematics, University of Belgrade, Serbia
E-mail: zarkom@matf.bg.ac.rs, nada@matf.bg.ac.rs, rviktor@matf.bg.ac.rs

Abstract. Even if A. Michelson, H. Lorentz, H. Poincaré and others performed experiments and wrote papers to the subject known later as the special theory of relativity more than two decades before Albert Einstein published his paper on this topic in 1905, it is usually considered that special theory of relativity originated in this year. This theory was soon accepted and many authors wrote papers on this matter. It is not so well known that several Serbian mathematicians and astronomers also wrote papers on special theory of relativity already in the twenties of XX century. Some of the papers were popular explanation of basics of this theory and they were intended to the general public, but there were also papers with original contributions. All papers were written in Serbian and certainly this is the main reason that these results were neglected. We shall present works in this field of a great Serbian astronomer and climatologist Milutin Milanković (1879-1958), a great Serbian mathematician and the founder of the Serbian mathematical school Mihailo Petrović Alas (1868-1943) and Sima Marković (1888-1939), mathematician, philosopher and politician, also known as one of the founder of the Communist Party of Yugoslavia.

1. INTRODUCTION

Albert Michelson, Hendrik Lorentz, Henri Poincaré and others performed experiments and wrote papers on the subject known later as the special theory of relativity in the interval ranging more than two decades before Albert Einstein published his first paper on this topic in 1905. However, it is usually considered that special theory of relativity originated in this year. A decade later the general theory of relativity was founded in the works of David Hilbert and Albert Einstein. These theories were soon accepted and many authors wrote papers on this matter. It is not so well known that several Serbian mathematicians and astronomers also wrote papers on the theory of relativity already in the twenties of XX century. Some of the papers were popular explanation of basics of these theories and they were intended to the general public, but there were also papers with original contributions. All papers were written in Serbian and certainly this is the main reason that results offered there were neglected. We shall present works in this field of a great Serbian mathematician and the founder of the Serbian mathematical school Mihailo Petrović Alas (1868-1943), the great Serbian astronomer and climatologist Milutin Milanković (1879-1958), and Sima Marković (1888-1939), the mathematician, philosopher and politician (Figure 1).

There are some common details in personalities and writings of all three authors. Their first characteristic is the universality. In terms of scientific work, all three belong to a specific time. Due to the accumulated scope of scientific and mathematical knowledge, it was difficult, if not impossible, for an individual to know the whole body of natural sciences and mathematics. The time of universal mathematicians and scientists was slowly passing by. By understanding and the wideness of scientific work in mathematics, mechanics and philosophy, Henri Poincaré, French mechanist and mathematician, was certainly one of the last homo universal of sciences. And it was Poincaré who was one of the professors of Mihailo Petrović. After a later work in science, we can conclude that the spirit of universalism of his professor passed on to Petrović. Sima Marković was a Petrović's student and this universal determination and the wide scientific curiosity certainly was transferred from Professor to his student. A similar characteristic feature was presented in the work of Milutin Milanković. Hence all of them were not working only in their basic fields, but in many other areas as well, including theory of relativity. Their second characteristic is associated to the relationship of mathematics and physics. All three had a point of view that both disciplines are highly interconnected. Petrović even invented the whole new philosophical system, mathematical phenomenology, in which the main point in science should have establishing analogies, including physically disparate phenomena, by reducing to the same abstract essence. Of course, mathematics must have the main role in this reduction, in Petrović's case that were differential equations. The greatest scientific achievement of Milutin Milanković was insolation theory which characterizes the climates of all the planets of the Solar system and explanation of Earth's long-term climate changes caused by changes in position of the Earth in comparison to the Sun. The main tool in the development of his theories had advanced mathematical method and equations of celestial mechanics, the subject which was considered then as applied mathematics. It took several decades to find empirical evidences which led to recognition and the final acceptance of his theory. As we shall see, Marković had a similar, or even stronger attitude to the ties of physics and mathematics. All of them were very good popularizers of science and that was their third characteristic. They wrote many articles and books which appreciate science, ranging from history and philosophy of science to ichthyology and travelogues. They were very good writers and some of their books are still very popular and a part of obligatory school reading. Just to mention Milanković's Trough distant worlds and times (Kroz vasionu i vekove), which consists of 37 letters on science addressed to an unnamed yang lady and the Petrović's books In the pirate's country (U zemlji gusara) and Eel's story (Roman jegulje).

Figure 1: From left to right: Mihailo Petrović, Sima Marković and Milutin Milanković.

Their writings on relativity appeared in the early stage of development of this theory. When their books appeared, there were discoveries that were still waiting to emerge. For example, Petrović's writing was printed in 1921, while Friedmann equations appeared in 1922. Marković's book was published in 1924, several years before Georges Lemaître and Edwin Hubble discovered the Universe expansion and the property of the universe what is now known as the Hubble law. They had a positive attitude to the theory of relativity and were very curious about it. But this not was the case in general. There were many opponents in science and even more in broad parts of general community, sometimes very hostile. To illustrate these circumstances, we mention the following extreme opinion of Justin Popović (1894-1979), Eastern Orthodox theologian, archimandrite and professor of Belgrade University (at the Faculty of Theology): "No doubt, relativism is also the logic, and the nature and soul of humanism. Einstein's theory of relativity is the final, aggregate resultant of humanism and all its philosophical, scientific technical and political branches. But not only this, nevertheless in the last line to its own, humanism is nothing else but - nihilism... Humanism inevitably evolves into atheism, passes through anarchism and ends with nihilism. If anyone today is an atheist, know, if he is consistent, tomorrow he will be an anarchist, and the day after tomorrow, a nihilist. And who is a nihilist, know, that it has come to him from humanism through atheism... Humanism is in fact the basic evil, the original evil of man." Justin Popović was proclaimed a saint in 2010.

2. MIHAILO PETROVIĆ ALAS

Mihailo Petrović was an outstanding mathematician and a colorful character of the public life of Belgrade in the first half of the twentieth century. He studied mathematics in Belgrade and Paris at Sorbonne (École normale supérieure) where he got diplomas in mathematics and physics. In the year 1894 he defended his doctoral dissertation on differential equations under supervision of the prominent mathematicians Charles Hermite and Charles Émile Picard. In the same year he returned to Belgrade and was elected for the professor of mathematics at the Grand School, the later University of Belgrade. He was scientifically very productive and soon became the leading Serbian mathematician. When he was only 32, he was elected as the full member of the Serbian Academy of Sciences. As the representative of French mathematical school, he brought in Belgrade the spirit of the contemporary mathematics, particularly in analysis and ordinary differential equations. He has the great merit for the development of mathematics in Serbia and by the general opinion, he is considered as the founder of Serbian mathematical school. For example he was the supervisor for all doctoral dissertations in Serbia until the Second World War and the founder of the first international Serbian mathematical journal.

Mihailo Petrović had a rich, interesting and an unconventional life. It is even difficult to list in one place everything that Petrović was dealing with, let alone to describe it. In addition to the interest in various branches in mathematics, we meet him in many other, often unexpected places. He was a writer of laws and proposals of interstate agreements, but also an inventor and owner of successful and realized patents. Petrović is the creator of an original theory in natural philosophy, mathematical phenomenology and by many he is included in the group of Serbian most important philosophers. He wrote several nice and interesting novels and some of them are a favorite part of Serbian youth literature. He wrote scientific papers and was interested in other natural sciences, e.g. in ichthyology, astronomy, theory of relativity, and chemistry. He was the creator of the coding system and the main cryptographer of the Serbian and later of Yugoslav army. He played violin and led the music band "Suz", which had one of the main places in the bohemian life of Belgrade. Finally, he was a big and passionate fisherman and a great world traveler and sailor in the north and south seas.

Petrović wrote the first paper on relativity under the title "Theory of relativity" already in 1921. The paper had 12 pages and was published in somewhat unusual journal regarding its content, Serbian literary messenger. But this is understandable if one knows that the paper was popular in nature and intended for the general audience. There were not formulas and most of the paper is devoted to special theory of relativity. The significant part of the paper is devoted to the genesis of this theory. Beside the explanation of the basics, he mentioned contributions of various people to this theory, e.g. Michelson - Morley experiment, then Lorentz - FitzGerald contraction, Poincare and Einstein final foundation of the theory. At the end he mentioned some phenomena related to the general theory of relativity, Mercury orbit precession and the light bending in the gravitational field. In this contexts he mentioned two expeditions to Principe (the northern island of São Tomé and Príncipe) and Sorbel (Brazil) organized respectively by Eddington and Crommelin in 1919. These famous expeditions had the goal to test Einstein prediction, the stars rays bending during the Solar eclipse.

The second paper, "The physical constants in the relativity theory" was published in Глас of the Serbian academy of science in 1927. The paper had 16 pages and was dealing with the measurement of time. The basic idea of the paper is "tacit assumptions", as Petrović wrote, that the physical constant, such as coefficient of electrical conductivity, are invariant from transitions from one inertial coordinate system to the another one. Petrović recalls a device and an experiment produced in 1887 by Gabriel Lippmann which he made for time measuring. The whole idea was that a certain interval time Δt is proportional to the electrical resistance ρ of mercury and that Δt can be measured by Lippman's device. Petrović discussed there two possibilities. The first conclusion was that the measurement of ρ and ρ' in two relativistic systems S and S' would depend not only on the speed of S and S' but also on the orientation of apparatus and even their design. If the absoluteness of ρ is assumed, i.e. $\rho = \rho'$ then we would have $\Delta t = \Delta t'$ what, of course, contradicts the special theory of relativity. Petrović's conclusion was that according to the theory of relativity physical constants do not have always the same values, they depend on how they are measured and they may be accidental. He suggested how to overcome this antinomy by proposing general forms of transformation for values of physical constants in two relativistic system. Petrović most probably met Lippman and was introduced in his ideas during his studies in Sorbonne since Lippmann there was professor of mathematics (since 1883) and physics (since 1886). Finally, we should mention in the favor of Petrovi's paper that the transformation rules for physical values in two relativistic systems in general is still open. Probably the most known example of this type is the relativistic transformation of temperature. M. Planck proposed a transformation formula in 1908, but H. Ott corrected it in the quite opposite way in 1963 (see H. Ott: Zeits. Phys. (1963) 175, 70).

Petrović's third paper on relativity was Physical etalons for time measuring, see Etalons physiques de temps, Publ. de l'Observatorie astronomiquc de l'Université de Belgrade, 1933, t. 11, pp. 5-10. The paper was published in 1933 and in the greatest part it is a compilation of his previous paper on this matter.

3. SIMA MARKOVIĆ

Sima Marković was a talented mathematician, gifted philosopher and a tragic figure at the Yugoslavian political scene and of the communist movement. He finished studies in mathematics with honor in 1911 and immediately became an assistant of Professor Petrović at the mathematical department of the Belgrade University. He completed his doctoral dissertation on Riccati differential equation in 1913, but the beginning of the First World War broke the academic activities in Serbia and his scientific work as well. After the war he started his political career, was a member of the parliament and in 1920 he was elected together with Filip Filipović, also a mathematician, for the first general secretary of Yugoslav communist party. In the same year the University nominated him for the assistant professor but due to his political activities, the Yugoslav king Alexandar did not want to sign the proclamation for this appointment. Soon he was expelled from the university and even more during twenties he was sent to prison several times where he spent altogether four years. In the thirties he moved to Moscow where he was the scientific associate at the Philosophical institute of the USSR Academy of Sciences, but had high positions in the communist establishment, too. He was also involved in the fractional fights in the party. Seemingly, he was on the wrong side as in 1939, April 19th, in Stalin's purges he was sentenced to death by shooting. He was executed in the same day. Ironically, he did not support offensive or violent methods in politics as he believed in means of parliamentary democracy, in the institutional and legal methods.

Marković's main interest in mathematics was teaching, education and methodic and he wrote several works on these topics. However, his main concern was the political philosophy and to some extent philosophy of science, particularly views based on physics. He wrote several books and booklets on the philosophy of science and related topics. For example, in the book The Principle of Causality and Modern Physics, Marković emphasizes the connection between physics and mathematics. He believed that the key feature of modern physics is its mathematics and that mathematics as all other sciences originates from the experience and from the needs of the everyday life. He also wrote in twenties a short book (80 pages) on the theory of relativity and a longer paper (31 pages) on the same topic published five years later. It interesting that he wrote the book during his stay in prison. The paper is more or less a compilation of the book, so we shall give some comments only on the book.

Sima Marković published his first work on relativity, the book Теорија релативитета, популарна-научна скица (Theory of relativity, a popular-scientific sketch) in Belgrade in 1924 (Figure 2). As the title says, the book is intended to the broad audience, particularly to those readers who are inclined to physics and science in general. It is interesting that the author in the foreword writes: "Even if Theory of relativity is physical by content, by form it still belongs to the most abstract parts of Theoretical mathematics". He glorifies mathematics often and compare it to the highest men's achievements. However, he did not use in the book a mathematical language, except few times some very simple formulas. He explains the main goal of the book as follows: "In this sketch, only the most important facts that characterize the theory of relativity are exposed to the goal to which it has to serve. This sketch represents, therefore, only the skeleton of Relativity Theory, which cannot be revived without mathematics. But, I hope, more careful readers can still get at least an approximate idea of what the Relativity Theory is.". The book consists of two parts:

- Part 1. Special theory of relativity, pages 6 - 45.
- Part 2. General theory of relativity, pages 46 -80.

The first part starts with presentation of views of classical mechanics, Galileo relativity principle and Newton's cosmology: absolute space and absolute time. We could say that Introduction does not differ much from those in modern books, even if the book was written almost a century ago. Rest of the Part 1 is devoted to the explanation of basics of special theory relativity. Relativity of length and time are explained and shown why in this theory is not possible to talk about simultaneous events. Lorentz coordinate transformations, which he called Lorentz formulas, and the contraction factor γ are explained. He presents how special theory of relativity succeeded in the explanation of some old problems in astronomy such as the stellar aberration is. He also explained four-vectors, namely events in the Minkowski space. He ends this part with the very modern view that the physical laws are nothing else but the invariants under coordinate transformation.

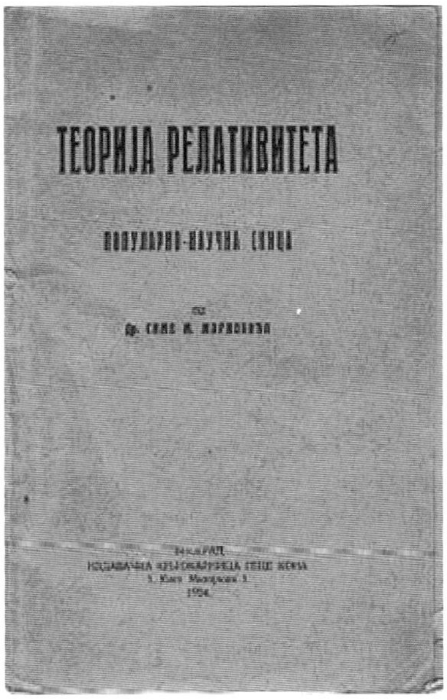

Figure 2: Cover page of the Theory of relativity, a popular-scientific sketch book by Sima Marković.

The second part starts with the discussion of the equivalence principle which deals with gravitational and inertial mass and the role of gravitation in General theory of relativity. There are presented virtually then all known facts from this theory. He mentioned main tests of General theory of relativity such as perihelion precession of Mercury's orbit and the bending of light rays in the gravitational fields. There is mentioned the Eddington's measure of star light bending during the Solar eclipse in 1919, just few years before the book appeared. He implicitly speaks about Riemann space explaining that General theory of relativity assumes a non Euclidian geometry. Some elements of differential geometry are mentioned such as the geodesics and the parallel transport needed for covariant derivative. It must be said that Marković showed the great skill in presenting these highly mathematized content, but not using a single mathematical symbol. He also described a phenomena such as the gravitational time dilation. He discussed also relation between the general theory of relativity and cosmology. He explains that the universe occupies a closed but unbounded space. As an application of the theory he deduced a formula of the size of the universe which depends only on the density ρ of the matter in the universe.

There are no references in the book, but the author mentioned that his book is based on the works of Einstein, Lorentz, Minkowski, Weyl, Riemann, Pauli, Helmholtz and few other physicist. The book is written in a nice and vivid language, and even today it could be interesting to a general reader.

4. MILUTIN MILANKOVIĆ

Milutin Milanković and his work are well known to astronomers. So we will highlight here just few facts from his life and work. Milanković is one of the Serbian greatest and most cited scientists of all times. His theory on the ice ages that accurately explains the change of climate on a large time scale is accepted world-wide. In recognition of his scientific achievements, a crater on the Moon, another one on the Mars and a planetoid were named after him. Besides his the most famous work Kanon der Erdbestrahlung und seine Anwendung auf das Eiszeitenproblem (published in 1941) and many other scientific papers, he also wrote an excellent book on celestial mechanics and books that popularize science. He is also known as a good civil engineer. Milanković came to Belgrade from Vienna in 1909 to teach applied mathematics at the Faculty of Philosophy of Belgrade University. His coming was a merit of Bogdan Gavrilović and Mihailo Petrović, who both taught mathematics at the University. Milanković was at the Belgrade University till his retirement in 1955, lecturing there applied mathematics. It should be mentioned that theoretical physics, mechanics and astronomy were considered then as the areas of applied mathematics. Milanković also was the first professor who started lecturing celestial mechanics at the University of Belgrade.

Milanković did two short excursions to relativity. He wrote his first paper "O teoriji Michelsonova eksperimenta" (On the theory of Michelson's experiment), Rad JAZU, vol. 190 pp 65-70 already in 1912. He was doing research in this theory in the period 1912-1924. In fact his papers on this matter were on special relativity and both are on Michelson experiment (now known as Michelson-Morley experiment) which gave the strong evidence against ether theory. He discussed there, in the light of the Michelson experiment, the validity of the second postulate of Special theory of relativity, that the speed of light is the same in all reference frames. Tatomir Anđelić and Andrija Stojković wrote an extensive paper on Milanković's views expressed in theses papers, so we shall not enter into discussion of his works on this matter. We should just mention that he was teaching special theory of relativity, most probably in arrangement with Mihailo Petrović, at the Belgrade university in the twenties of the last century.

5. OTHER SERBIAN SCIENTISTS

There were few other Serbian scientists who were engaged in scientific work or wrote about the theory of relativity. Probably Mileva Einstein - Marić (1875-1948), a mathematician and the first Einstein's spouse, is the most known as a possible contributor to the theory of relativity. There is a big debate if she really had any major contribution to the early Einstein work. We will not enter in this discussion, mainly because we do not know her signed scientific papers. Vladimir Varićak (1865-1942), the well-known mathematician of the Serbian origin, was a professor at the University of Zagreb. He wrote first of all, already in 1910, a paper on special theory of relativity. It interesting that he was a Milanković's high-school teacher. Few physicists and philosophers, Milivoje Dobrosavljević, Ognjen Prica and Dušan Nedeljković also discussed and wrote before the Second World War on the theory of relativity in various contexts. Nikola Tesla, the famous inventers, was strongly against this theory, while Mihailo Pupin, the other famous Serbian electrical engineer ant the professor at the Columbia University in New York, accepted it (Anđelić, Stojković, 1983).

References

Anđelić, T., Stojković, A.: 1983, „Milutin Milankovic et la theorié de la relativité“, Extrait du Rec. de textes, contrib. A la philosophie et la science a l'occasion du 80 anniv. de l'acad, Dusan Nedeljkovic, Ed. spec. Acad. Serbe des Sci. et des Arts, vol. DLI, Classe: des sciences sociales, vol. 91.

Mareš, J.J., Hubík, P., Šesták, J., Špička, V., Krištofik, J., Stávek, J.: 2016, „Relativistic transformation of temperature and Mosengeil-Ott's antinomy, arXiv:1606.02127 [physics.class-ph].

Marković, S.: 1924, *Teorija relativiteta*, Knjižnica Gece Kona.

Pejović, N., Malkov, S., Mitić, N., Mijajlović, Ž.: 2015, „Scientific papers of Milutin Milanković in his Digital Legacy“, *NCD Review*, **26**, 69–76.

Petrović, M.: 1933, „Étalons physiques de temps“, *Publ de l'Observatoire astronomiquc de l'Universitc de Belgrade*, **11**, 5-10.

Sources

Vrtual Library, Faculty of Mathematics, Belgrade: http://elibrary.matf.bg.ac.rs
Digital legacies, Faculty of Mathematics, Belgrade: http://legati.matf.bg.ac.rs

Proceedings of the XI Bulgarian-Serbian Astronomical Conference
(XI BSAC) Belogradchik, Bulgaria, May 14-18, 2018
Editors: Milcho K. Tsvetkov, Milan S. Dimitrijević and Momchil Dechev
Publ. Astron. Soc. "Rudjer Bošković" No 18, 2018, 109-116

RECTANGULAR TORUS DYNAMO MODEL AND MAGNETIC FIELDS IN THE OUTER RINGS OF GALAXIES

EVGENY MIKHAILOV and ALENA KHOKHRYAKOVA

Faculty of Physics, M.V.Lomonosov Moscow State University
Leninskie gori 1, 119991 Moscow, Russia
E-mail: ea.mikhajlov@physics.msu.ru

Abstract. Now it is no doubt that some spiral galaxies have regular magnetic fields of several microgauss. Their generation is described by the dynamo mechanism based on joint action of differential rotation and alpha-effect. For most of the galaxies the field generation is described by no-z model, which takes into account that the disc is quite thin. Some of the galaxies have outer rings, where we can suppose the magnetic fields, too. As for these objects, it is more useful to take torus dynamo model, which takes into account the vertical field structure. We describe the field evolution with both of these models for typical values of the parameters. It is quite interesting that the torus dynamo model can give dipolar magnetic fields (no-z model gives the quadrupolar field only).

1. INTRODUCTION

Nowadays it is well-known that several spiral galaxies have the magnetic fields of order of microgauss. Their existence is strongly proved observationally while studying the Faraday rotation measure for the radiowaves. The research of the magnetic fields is done by such modern radio interferometers as LOFAR, VLA, SKA (in future). The magnetic fields are also confirmed by synchrotron emission and another effects (Beck et al 1996; Arshakian et al 2009).

From the theoretical point of view, the magnetic fields are described by the galaxy dynamo mechanism. It is connected with the transition of energy of turbulent motions to the energy of the magnetic fields. It is based on joint action of two different effects. The first one is the differential rotation (characterizng the non-solid rotation of the galaxy). It transforms the radial component of the magnetic field to the angular one. The second part of the dynamo mechanism is alpha-effect, which describes the vorticity of the turbulent motions of the interstellar medium. The alpha-effect transforms the angular magnetic field to the radial one. They compete with the turbulent diffusion which makes the field decay. So the magnetic field generation is a threshold mechanism. If the turbulent motions of the interstellar medium are quite intensive, the magnetic field grows, else the large-scale structures of the field are destroyed.

The dynamo mechanism is usually described by Steenbeck – Krause – Raedler equations (Steenbeck et al 1966). They include the angular velocity of the galaxy and the alpha-effect coefficient. Unfortunately, the equations for the magnetic field are usually quite difficult both for theoretical and numerical analysis, and they are solved using different approximation. One of the most popular models is the no-z approximation which is based on the fact that the galaxy disc is very thin, and the ratio between half-thickness and radius can be assumed as a small parameter (Subramanian & Mestel 1993; Moss 1995; Phillips 2001). So we can assume that the field lies in the equatorial plane. For the derivatives of the field we use algebraic expressions, which allow us to use 2-dimensional model (which can be reduced even to spatially 1-dimensional system for the axisymmetric case).

Some of the galaxies have so-called outer rings, which are situated at some distance from the main disc. The observations show that there is ionized medium in such objects, also they demonstrate the alpha-effect and differential rotation there. So it is quite normal to apologize the magnetic field existence there (Moss et al 2016). As for the models of the magnetic field, we can use no-*z* approximation, too. However, for the outer ring the typical ratio between the half-thickness of the ring and its half-width cannot be used as a small parameter. So such calculations can give only qualitative results.

Another opportunity is given by the torus dynamo model (Deinzer et al 1993; Brooke & Moss 1992; Mikhailov 2017). It does not assume that the vertical component of the field is negligible, and uses different model assumptions. The field is divided to the angular part and the part which is described by the angular component of the vector potential of the magnetic field. This model describes the field more precisely. Another advantage of the torus dynamo model is that it can describe the field with different structures. The no-*z* model can give only quadrupolar field configurations and the torus model can describe dipolar magnetic fields, too.

We should also describe non-linear saturation of the field growth. The magnetic field cannot increase infinitely. The field evolution can be restricted from energetic arguments. The field growth is connected with the transition of the energy of the turbulent motions to the field energy. So if the field reaches so-called equipartition value, the growth should stop. The nonlinear effects can cause some important features of the field evolution, mainly for the dipolar one.

Here we present the results of modeling the magnetic fields in the outer rings using both the no-z approximation and the torus dynamo model. We compare the typical field configurations for different cases.

2. BASIC EQUATIONS

The magnetic fields of galaxies contain two different components. The first one is random (or small-scale – it has typical length-scale of several tens parsecs) and the second one is obtained as a result of averaging the field in the regions of 50 – 100 pc. Here we will describe the regular component of the magnetic field, which is described by the Steenbeck – Krause – Raedler equations (Steenbeck et al 1971):

$$\frac{\partial \vec{B}}{\partial t} = \mathrm{curl}\,\left[\vec{V}, \vec{B}\right] + \mathrm{curl}\,\left(\alpha \vec{B}\right) + \eta \Delta \vec{B};$$

where $\vec{B}$ is the large-scale magnetic field, $\vec{V}$ is the velocity of the large-scale motions, α characterizes alpha-effect and η is the turbulent diffusivity coefficient. For the velocity we assume that it is connected with the rotation of the galaxy:

$$\vec{V} = r\Omega\,\vec{e}_{\varphi};$$

where r is the distance from galaxy center and Ω is the angular velocity of the rotation. For the angular velocity we can take the model:

$$\frac{d\Omega}{dr} = -\frac{\Omega}{r}.$$

As for the alpha-effect we can use the following simple model (Arshakian et al 2009):

$$\alpha_0 = \frac{\Omega l^2}{h},$$

where l is the lengthscale of turbulence, and h is the half-thickness of the galaxy disc.

For the turbulent diffusivity we can take that:

$$\eta = \frac{lv}{3},$$

where v is the typical turbulent velocity.

We will assume the field for the following values of the parameters:

$$R - a < r < R + a;$$

$$\frac{a}{k} < z < \frac{a}{k}.$$

This vector equation is a system of three spatially three-dimensional equations. This fact makes quite difficult the possible analysis of the field growth. So we should use some approximations to make the study more simple.

3. NO-Z APPROXIMATION

If we take into account that the field nearly lies in the equatorial plane, we can use the next model of the horizontal components of the field:

$$B_r(r,\varphi,z,t)=B_r(r,\varphi,0,t)\cos\left(\frac{\pi z}{2h}\right);$$
$$B_\varphi(r,\varphi,z,t)=B_\varphi(r,\varphi,0,t)\cos\left(\frac{\pi z}{2h}\right).$$

So the field is assumed quadrupolar. We also shall assume the field axisymmetric (which is quite normal for the outer ring lying in the equatorial plane).

For the alpha-effect we will take the model:

$$\alpha=\alpha_0\frac{z}{h},$$

So we shall have the next equations for the magnetic field (Moss 1995):

$$\frac{\partial B_r}{\partial t}=-\frac{\Omega l^2}{h^2}B_\varphi-\eta\frac{\pi^2k^2B_r}{a^2}+\eta\frac{\partial}{\partial r}\left(\frac{1}{r}\frac{\partial}{\partial r}(rB_r)\right);$$
$$\frac{\partial B_\varphi}{\partial t}=-\Omega B_\kappa-\eta\frac{\pi^2k^2B_\varphi}{a^2}+\eta\frac{\partial}{\partial r}\left(\frac{1}{r}\frac{\partial}{\partial r}(rB_\varphi)\right).$$

The field evolution is restricted by the so-called equipartirion field. If the energy density of the magnetic field becomes equal the density of the turbulent motions energy (Arshakian et al 2009):

$$\frac{B_{\max}^2}{8\pi}=\frac{\rho v^2}{2},$$

the field growth should stop. This can be taken into account using the following modification of the equations:

$$\frac{\partial B_r}{\partial t}=-\frac{\Omega l^2}{h^2}B_\varphi\left(1-\frac{B_r^2+B_\varphi^2}{B_{\max}^2}\right)-\eta\frac{\pi^2k^2B_r}{a^2}+\eta\frac{\partial}{\partial r}\left(\frac{1}{r}\frac{\partial}{\partial r}(rB_r)\right);$$
$$\frac{\partial B_\varphi}{\partial t}=-\Omega B_\kappa-\eta\frac{\pi^2k^2B_\varphi}{a^2}+\eta\frac{\partial}{\partial r}\left(\frac{1}{r}\frac{\partial}{\partial r}(rB_\varphi)\right).$$

We can measure time in units of a^2/η, distances in R, and the magnetic field in equipartition units. So the equations for the field will be the following:

$$\frac{\partial B_r}{\partial t}=-S_\alpha B_\varphi\left(1-B_r^2-B_\varphi^2\right)-\frac{\pi^2k^2B_r}{4}+\lambda^2\frac{\partial}{\partial r}\left(\frac{1}{r}\frac{\partial}{\partial r}(rB_r)\right);$$
$$\frac{\partial B_\varphi}{\partial t}=-S_\omega B_\kappa-\frac{\pi^2k^2B_\varphi}{4}+\lambda^2\frac{\partial}{\partial r}\left(\frac{1}{r}\frac{\partial}{\partial r}(rB_\varphi)\right);$$

where S_α characterizes alpha-effect, S_ω describes differential rotation and $\lambda=\frac{a}{R}$ describes the half-width of the outer ring. For this case we should study the field with following boundary conditions:

$$B_r\Big|_{r=1-\lambda}=B_r\Big|_{1+\lambda}=B_\varphi\Big|_{1-\lambda}=B_\varphi\Big|_{1+\lambda}=0.$$

The field evolution will be described by the dimensionless number $Q = S_{\alpha} S_{\omega}$.
We describe the results for this case on figure 1. The field for this case is strongly quadrupolar.

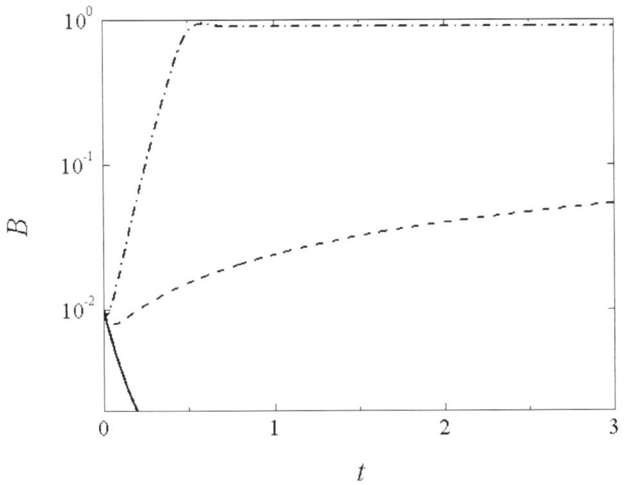

Figure 1: Magnetic field evolution described by no-*z* approximation. Solid line shows *Q=10*, dashed line – *Q=150*, dot-dashed line – *Q=500*.

3. TORUS DYNAMO MODEL

As we have said, it is much more convenient to use the torus dynamo model for the outer ring (Mikhailov 2017; Mikhailov 2018). The field is assumed as a combination of two different components:

$$\vec{B} = B\vec{e}_{\varphi} + \text{curl}\left(A\vec{e}_{\varphi}\right),$$

where B is the angular component of the magnetic field, and A is the angular component of the vector potential.

In the axisymmetric case we shall have the following equations (in nonlinear case):

$$\frac{\partial A}{\partial t} = -\frac{\Omega l^2}{h^2} B\left(1 - \frac{B^2}{B_{\max}^2}\right) + \eta \Delta A;$$

$$\frac{\partial B}{\partial t} = \Omega \frac{\partial A}{\partial z} + \eta \Delta B.$$

Using the same dimensionless numbers, we can obtain the equations:

$$\frac{\partial A}{\partial t} = -S_{\alpha} B\left(1 - B^2\right) + \eta \Delta A;$$

$$\frac{\partial B}{\partial t} = S_{\omega} \frac{\partial A}{\partial z} + \eta \Delta B.$$

The field for different Q ($k=2$) is shown on figure 2. It is quite important, that for high D the magnetic field can have quadrupolar symmetry. The field structure for quadrupolar case ($D=150$) is shown on figure 3 – 4. The dipolar field (D=500) is presented on figure 5 – 6.

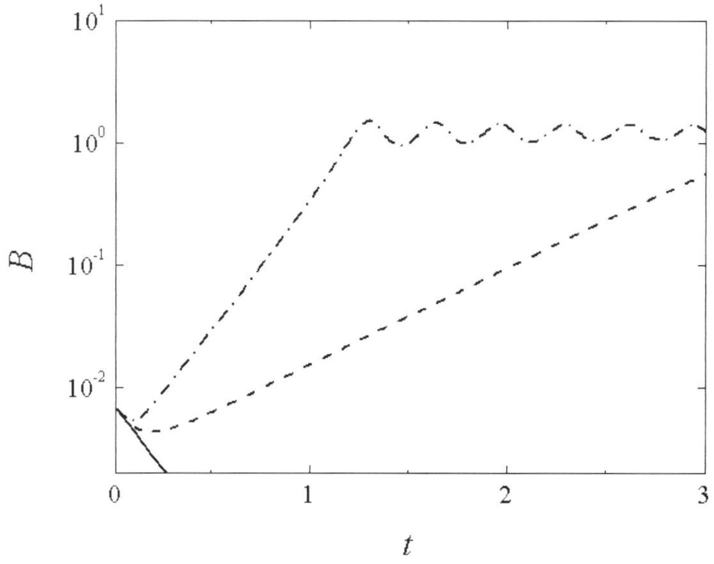

Figure 2: Magnetic field evolution described by torus dynamo model. Solid line shows $Q=10$, dashed line – $Q=150$, dot-dashed line – $Q=500$.

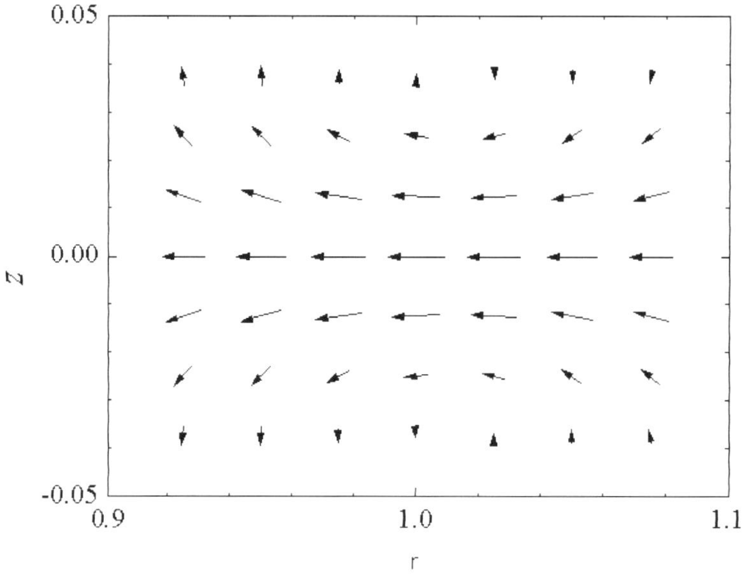

Figure 3: Poloidal magnetic field structure for $Q=150$ (quadrupolar field).

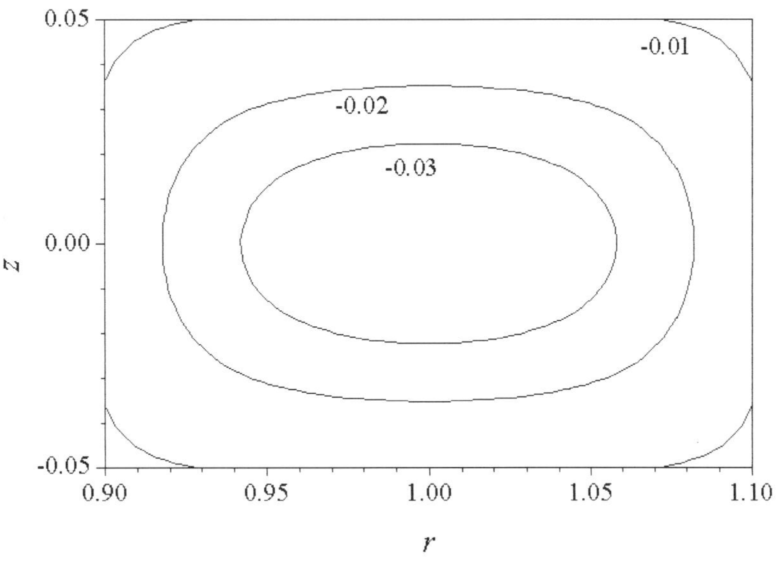

Figure 4: Toroidal magnetic field structure for *Q=150* (quadrupolar field).

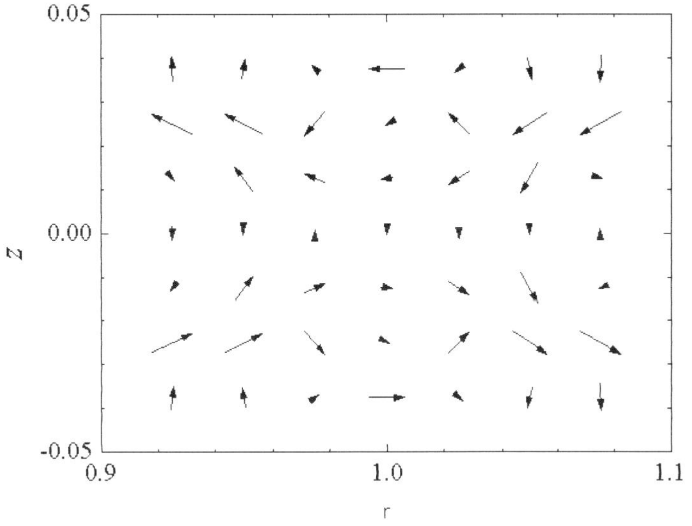

Figure 5: Poloidal magnetic field structure for *Q=500* (dipolar field).

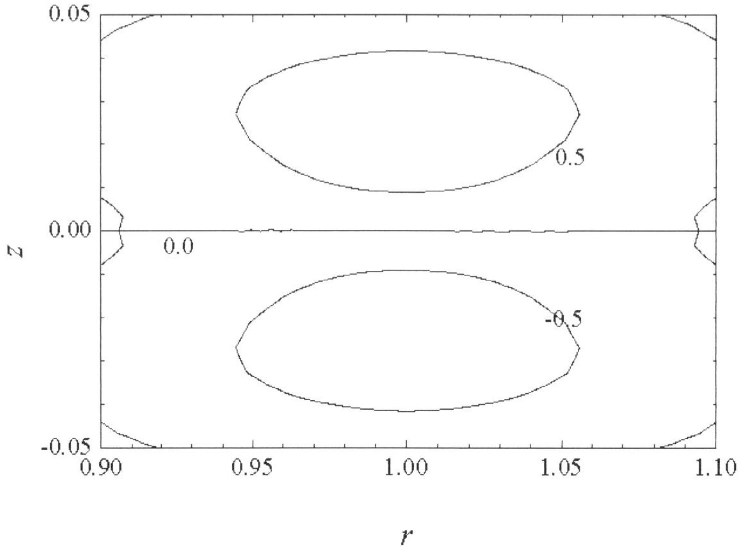

Figure 6: Toroidal magnetic field structure for *Q=500* (dipolar field).

4. CONCLUSIONS

We have studied the magnetic field in the outer rings using both no-*z* approximation and torus dynamo model. The second one is thought to be more realistic. The magnetic field grows slower than for first model. Also the magnetic field can have not only quadrupolar symmetry. If the kinetic energy of the turbulent motion is quite high (*D>153*), the field can become dipolar.

It could be quite important to study the magnetic fields in the outer rings observationally. There are only some observations for the outer ring of NGC4736. In principle, the results are quite correspond to our modelling, but it also would be important to stude some another objects.

This work was supported by RFBR (16-01-00447). We are also grateful to the organizers of XI Bulgarian – Serbian Astronomical Conference for an opportunity to present our work.

References

Beck, R., Brandenburg, A., Moss, D. et al. 1996, *Ann. Rev. Astron. Astrophys.*, **34**, 155 – 206.
Arshakian, T., Beck, R., Krause, M., Sokoloff, D.: 2009, *Astron. Astrophys.*, **494**, 21 - 32.
Steenbeck, M., Krause, F., Raedler, K.H. : 1971, *The Turbulent Dynamo. A translation of a series of papers by F. Krause, K.-H. Radler, and M. Steenbeck.*, 29 – 47.
Subramanian, K., Mestel, L.: 1993, *Mon. Not. R. Astr. Soc*, **265**, 649 – 654.
Moss, D. : 1995, *Mon. Not. R. Astr. Soc*, **275**, 191 – 194.
Phillips, A.: 2001, *Geophys. Astrophys. Fluid Dyn.*, **94**, 135 – 150.
Moss, D., Mikhailov, E., Silchenko, O. et al. 2016, *Astron. Astrophys.,* **592**, A44.
Deinzer, W., Grosser, H., Schmitt, D. : 1993, *Astron. Astrophys.*, **273**, 405 – 414.
Brooke, J. M., Moss, D. : 1994, *Mon. Not. R. Astr. Soc.,* **266**, 733 – 739.
Mikhailov, E. : 2017, *Astron. Rep.*, **61**, 739 – 746.
Mikhailov, E. : 2018, *Astrophysics,* **61**, 147 – 159.
Chyzy, K. T., Buta, R. J. : 2008, *Astrophys. J.*, **677**, L17 – L 20.

Proceedings of the XI Bulgarian-Serbian Astronomical Conference (XI BSAC) Belogradchik, Bulgaria, May 14-18, 2018
Editors: Milcho K. Tsvetkov, Milan S. Dimitrijević and Momchil Dechev,
Publ. Astron. Soc. "Rudjer Bošković" No 18, 2018, 117-125

HARD X-RAY DIAGNOSTIC OF PROTON PRODUCING SOLAR FLARES COMPARED TO OTHER EMISSION SIGNATURES

ROSITSA MITEVA[1], KOSTADINKA KOLEVA[2], MOMCHIL DECHEV[2], ASTRID VERONIG[3], KAMEN KOZAREV[2], MANUELA TEMMER[3], KARIN DISSAUER[3] and PETER DUCHLEV[2]

[1]*Space Research and Technology Institute, Bulgarian Academy of Sciences Acad. Georgi Bonchev Str., Block 1, 1113 Sofia, Bulgaria*
[2]*Institute of Astronomy with National Astronomical Observatory, 72 Tsarigradsko Chaussee Blvd., 1784 Sofia, Bulgaria*
[3]*Institute of Physics/IGAM, University of Graz, Universitätsplatz 5, A-8010 Graz, Austria*
E-mail: rmiteva@space.bas.bg, koleva@astro.bas.bg, mdechev@astro.bas.bg, astrid.veronig@uni-graz.at, kkozarev@astro.bas.bg, manuela.temmer@uni-graz.at, karin.dissauer@uni-graz.at, pduchlev@astro.bas.bg

Abstract. We present results on the correlation analysis between the peak intensity of the in situ proton events from SOHO/ERNE instrument and the properties of their solar origin, solar flares and coronal mass ejections (CMEs). Starting at the RHESSI mission launch after 2002, 70 flares well-observed in hard X-rays (HXRs) that are also accompanied with in situ proton events are selected. In addition to HXRs, flare emission at several other wavelengths, namely in the soft X-ray (SXR), ultraviolet (UV) and microwave (MW), is used. We calculated Pearson correlation coefficients between the proton peak intensities from one side, and, from another, the peak flare flux at various wavelengths or the speed of the accompanied CME. We obtain the highest correlations with the CME speed, with the SXR flare class and with MWs, lower ones with the SXR derivative, UV and 12–50 keV HXRs and the lowest correlation coefficients are obtained with the 50–300 keV HXRs. Possible interpretations are discussed.

1. INTRODUCTION

The quest towards the solar origin of in situ observed solar energetic particles (SEPs: electrons, protons and heavy ions at energies above keV, Desai and Giacalone, 2016) is still ongoing after multiple decades of research (Basilevskaya, 2017). From one side, the main mechanisms for particle acceleration out of thermal populations in the solar corona are well known: magnetic reconnection and shock waves (Klein and Dalla, 2017). The eruptive processes on the Sun where both of these acceleration processes take place are solar flares (SFs) and coronal mass ejections (CMEs). However, the individual contribution of SFs and CMEs to the in situ detected particle flux is unknown. The initial general consensus of predominant acceleration by flares later shifted to the expectation that CMEs cause most, if not all, of the particle acceleration (when CMEs were discovered). Further work led to the impulsive vs. gradual SEP classification doctrine (Reames 1999), but after several proposed modifications (e.g., Cliver 2009), consensus finally settled on the mixed-origin contribution standpoint (e.g., Cane et al. 2010 and references therein). Although, both, SFs and CMEs are nowadays accepted to accelerate SEPs, their individual contributions to the observed particle flux event are hard to separate. The standard procedure when dealing with a large sample of events is to calculate Pearson correlation coefficients between the peak proton flux (or an integrated value, fluence) and some characteristic value for the SF (i.e., flare class, fluence and rise time) and CME (i.e., linear speed and angular width), usually adopted from catalogs (see discussion in Dierckxsens et al. 2015). Several guiding principles must be considered in the SEP research, e.g., the big flare syndrome (Kahler 1982) and the longitudinal effect on the particle profiles and intensities (e.g., Lario et al. (2013) for multi-spacecraft observations).

Solar flare intensity is defined as the peak value (measured in W/m^2) in the 1–8 Å channel (i.e., soft X-rays, SXRs) observed by GOES satellites (https://satdat.ngdc.noaa.gov/sem/goes/data/). The wavelength is representative of thermal emission. SXR data is freely provided by the GOES team since the 1970s, and this energy channel is nowadays used for the definition of SF class (from so-called A (being a multiplication factor of 10^{-8}) to X (10^{-4}) flare class). The electromagnetic (EM) emission during (especially large) SFs could span from the gamma to radio range (Benz 2002, 2017, Fletcher 2011). The light-curve, depending on the wavelength, has specific time profile, duration, acceleration mechanism, emission site in the solar atmosphere, etc. In this study we use SXR, hard X-ray (HXR), ultraviolet (UV) and radio data.

With the launch of the RHESSI satellite in 2002 (until its decommissioning in 2018), HXRs and gamma-rays from solar flares have routinely been measured according to the observing mode of the spacecraft (Lin et al. 2002). The HXR instrument observes flare emission from 3 keV up to 20 MeV as counts per seconds. Different models can be subsequently applied to reconstruct the HXR photon flux from the measured count rates.

Previous studies relating HXRs and proton events are reported by Kiplinger (1995), with a focus on the spectral properties of the flares using Solar Maximum mission (SMM) data. The study started with selection of 152 HXR flares, well-observed with SMM, and explored their link to interplanetary proton events. The study concluded that absence of flare spectral hardening is associated (to over 95%) to the absence of proton events and vice versa (82%). In a second analysis,

193 less intense flares, compared to the first sample, were used to successfully predict occurrence of large proton events.

The recent state of the art instrument in the extreme- and ultraviolet (EUV and UV, respectively) range is the Atmospheric Imaging Assembly (AIA) aboard the Solar Dynamics Observatory (SDO; Lemen et al. 2012). AIA observes the Sun in 10 wavelengths with a cadence as high as 12 seconds.

In the radio domain we use data from the Radio Station Telescope Network (RSTN, Guidice et al. 1981) which is a set of four nearly identical stations (located in Learmonth – Western Australia, San Vito – Southern Italy, Sagamore Hill – Massachusetts, USA and Palehua – Hawaii, USA). The temporal coverage is nearly complete. Observations at eight discrete frequencies from 245 MHz to 15.4 GHz are routinely done with 1-second time resolution. Radio flux density is observed in solar flux units (sfu). For the purpose of this analysis (favoring signatures of non-thermal emission) we use the highest available frequency, 15.4 GHz, falling in the microwave (MW) range.

The aim of the present study is to statistically investigate new EM wavelengths as alternatives of the well-known SXR flux using Pearson correlation analysis widely used to quantify the solar origin of SEP events. Adopting HXRs as a reference emission, all other wavelengths are the EM signatures of the proton-producing HXR flares.

2. DATA AND ANALYSIS

For the current study we used in situ protons in the 17–22 (~20) MeV energy channel of SOHO/ERNE instrument (Torsti et al. 1995). The proton events are identified in the period 1996–2017 (Miteva and Danov 2018) and constitute the reference channel of an on-line catalog under development (Miteva 2017, Miteva and Danov 2017): **http://newserver.stil.bas.bg/SEPcatalog/**. About 660 proton events are identified at present. An example of the proton time profile is presented in Miteva (2017).

In addition, temporal and longitudinal association between the proton events and the solar eruptive phenomena, occurring prior the SEP increase at 1 AU, is performed. Namely a pair (where possible) of flare and CME are assigned as the most probable origin to each proton event (see Miteva 2017 for the details of the adopted procedure). The number of solar associations is always reduced with respect to the total number of identified proton enhancements.

The so-identified list of solar flares (close to 400 flares) is subsequently used to firstly check if the flare is observed by the RHESSI satellite using the RHESSI browser 2.0 tool (**http://sprg.ssl.berkeley.edu/~tohban/browser/**) – see examples in Fig. 1. Due to the rotation and orbit characteristics of the satellite, periods of observations are interrupted by night time and passage through the South Atlantic Anomaly. This fact reduced drastically the number of detected flare events. As additional condition, HXR data should be available during the observations at least from the onset to the peak of the emission at several HXR channels. Thus, a list of 70 events with protons and flare HXR emission is finally compiled in the present analysis.

Complementary to the RHESSI HXRs, we use the standard wavelength used in the definition for the flare class and provided by the GOES 1–8 Å SXR channel. In addition we use SDO/AIA 1600 Å channel (sensitive to upper photosphere and transition region) and RSTN 15.4 GHz microwaves (radio emission from the lower layers of the solar atmosphere). As an example, we show the light curves for the 15-Feb-2011 flare: X2.2 in SXR GOES class (provided by the quick-look RHESSI browser, Fig. 1 left), RHESSI HXRs (3–300 keV, Fig. 1 right), SDO/AIA UV (Fig. 2 left) and RSTN MWs (Fig. 2 right). SXR and UV are smooth peaks, whereas HXR and MW emission show more bursts.

Each proton event in our list is also linked to a CME (unless there is a data gap or the association is uncertain). For comparative purposes, we perform correlations between the proton peak intensity and the linear speed of the CMEs adopted here from the SOHO LASCO CME catalog (https://cdaw.gsfc.nasa.gov/CME_list/).

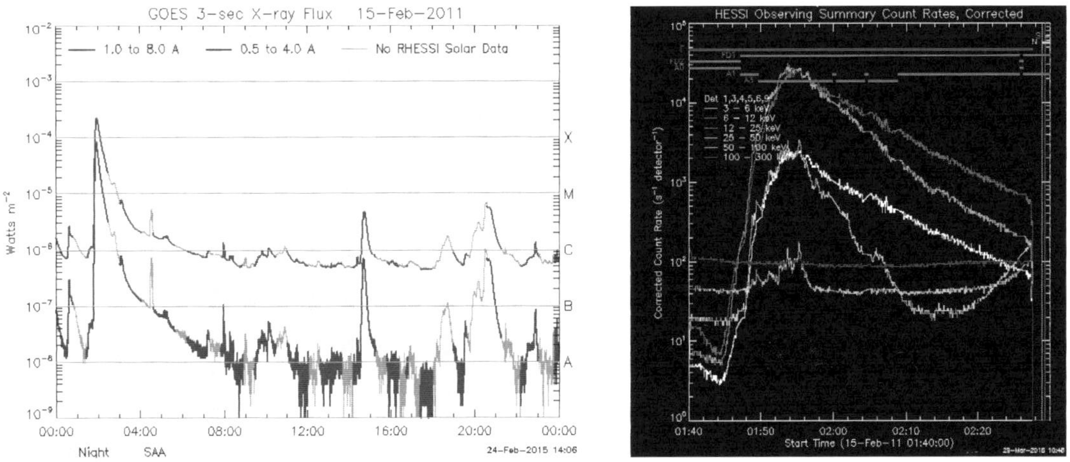

Figure 1: Example for 15-Feb-2011 solar flare observations in GOES SXRs (left) and RHESSI corrected count rates at several energy channels (right).

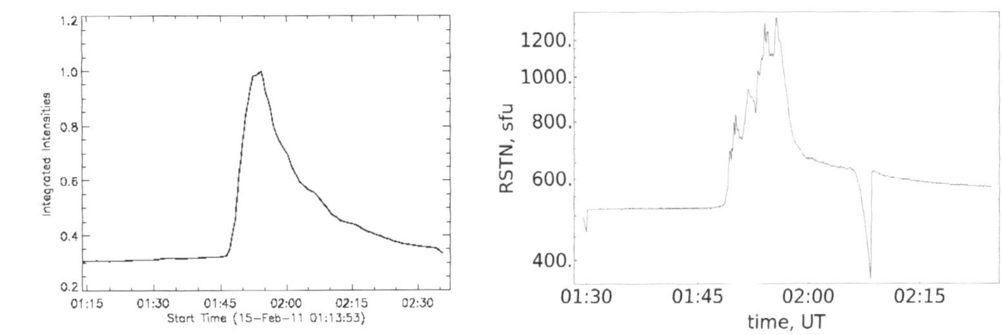

Figure 2: Examples for the 15-Feb-2011 flare light curves in 1600 Å UV (left) and 15.4 GHz MW (right) wavelengths. The abrupt drop in the declining MW phase is a data artifact.

3. RESULTS

The following procedure for the analysis of all EM curves (HXR, SXR, UV and MW) is performed. We first calculate background subtracted amplitudes. The analysis starts with a manual selection of the start and end times of the quiet-time intensity. The intensity over this period is averaged to give the (pre-event) background level for each flare. The latter value is subsequently subtracted from the identified peak value (i.e., from the highest intensity level of the light-curve). A similar procedure is applied also to the proton data to calculate the SEP proton peak fluxes.

The number of events used further are as follows: 70 events for 12–25 keV HXRs and SXR flare class; the SXR derivative could be calculated for all but one case; with the increase of the energy, the event sample is reduced to 64 events with HXR signal in 25–50 keV, 55 – in 50–100 keV to only 34 events with 100–300 keV HXRs. Similarly, radio data at 15.4 GHz is found for 50 events and UV data is available only for 22 events (mostly due to the SDO/AIA coverage).

Since the time derivative of SXR is commonly used as an approximation for the HXRs, so-called Neupert effect (Dennis and Zarro, 1993, Veronig et al. 2005), we performed a cross-correlation between the values for the GOES derivative and the RHESSI count rates in 4 different energy channels (Fig. 3). Pearson correlation coefficient is calculated, as well as an uncertainty using the bootstrapping method (Wall and Jenkins, 2003). The largest values for the correlations are obtained for the 'softest|' HXR 12–25 MeV (0.93±0.02) and 25–50 MeV (0.82±0.06) channels, being reduced greatly at the harder HXRs, 50–100 keV (0.57±0.12) and 100–300 keV (0.45±0.18). The event number for each calculation is limited to the sample size of the higher HXR energy in each pair.

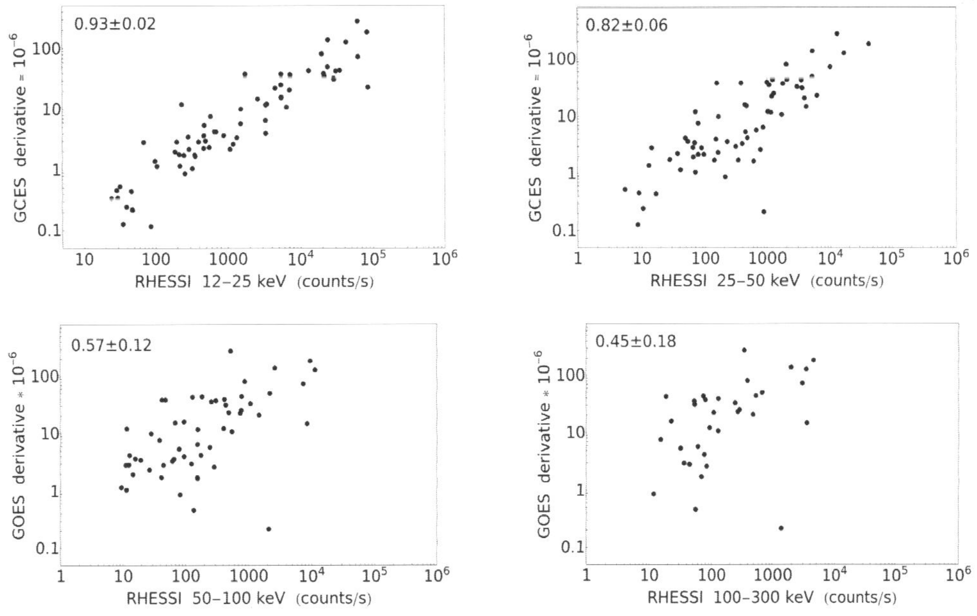

Figure 3: Scatter plots of GOES SXR derivative and RHESSI count rates in different energy channels.

Table 1: Correlation coefficients between the ~20 MeV SOHO/ERNE peak proton intensity and the properties of flare EM emission and CME speed. The number of events used in each calculation is given in brackets. See text for abbreviations used.

Solar origin properties	Correlation coefficients	
	All events	***Well-connected events***
Flare EM emission		
SXR 1–8 Å, W/m^2	0.56±0.09 (70)	0.61±0.09 (52)
SXR derivative, W/(m^2 s)	0.48±0.09 (69)	0.50±0.10 (52)
HXR 12–25 keV, counts/s	0.48±0.08 (70)	0.50±0.10 (51)
HXR 25–50 keV, counts/s	0.50±0.09 (64)	0.50±0.11 (47)
HXR 50–100 keV, counts/s	0.44±0.11 (55)	0.38±0.13 (41)
HXR 100–300 keV, counts/s	0.41±0.12 (34)	0.42±0.13 (28)
MW 15.4 GHz, sfu	0.55±0.10 (50)	0.62±0.11 (35)
UV 1600 Å, relative units	0.50±0.15 (22)	0.43±0.20 (15)
CME speed, km/s	**0.64±0.08 (65)**	**0.72±0.07 (50)**

The main objective of the analysis is the calculation of the Pearson correlation coefficients between the peak proton intensity at ~20 MeV and the EM emission amplitudes at each wavelength under consideration. For the purpose of comparison, we also used the speed of the CME. All results are summarized in Table. 1 and several of the examples are explicitly shown as scatter plots in Fig. 4.

We performed the log10–log10 correlations for the entire sample ('All events') and for a subset of events that fulfill the condition to originate from western helio-longitudes (or so-called here 'Well-connected events'). The sample is dominated by well-connected events (i.e., originating from western helio-longitudes, denoted with filled circles in Fig. 4, whereas the remaining cases are eastern events – plotted in open circles, respectively) which is a well-known property of SEP events. With the exception of two wavelengths (HXR 50–100 keV and UV 1600 Å) the correlation coefficients are higher (or the same, as for HXR 25–50 keV) for the well-connected events compared to the entire sample. Since the uncertainty ranges are fairly large, the improvement is, however, not statistically significant.

The standard correlation used to quantify the solar origin contribution to the proton flux is between the proton peak intensity with the SF class and/or with the CME linear speed. The correlations with the CMEs is slightly better (0.64±0.08) compared to the SXR flare class (0.56±0.09) considering the (nearly) entire event list. The same trend is kept also for the well-connected sample. Again, the differences are within the error bars.

The correlation coefficients obtained between the proton peak flux and SXR derivative, as well as between the proton flux and HXR peak value at 12–25 keV are very similar. When using the higher energy channel 25–50 keV the values are still very close to those with the SXR derivative. This is due to the highest cross-correlation between the two types of data (Fig. 3). Considering the highest energy channels used here (>50 keV) for the HXRs, we obtained smaller values for the correlation coefficients (the lowest for the 100–300 keV HXRs, 0.41±0.12), compared to the SXRs and statistically lower (especially for the western events) than those with the CME speed. The number of events at 100–300 keV is also reduced compared to the initial sample size, which also increases the uncertainty. The respective values for the MWs are very similar to the SXRs, (0.55±0.10), or slightly lower when using UV data (0.50±0.12).

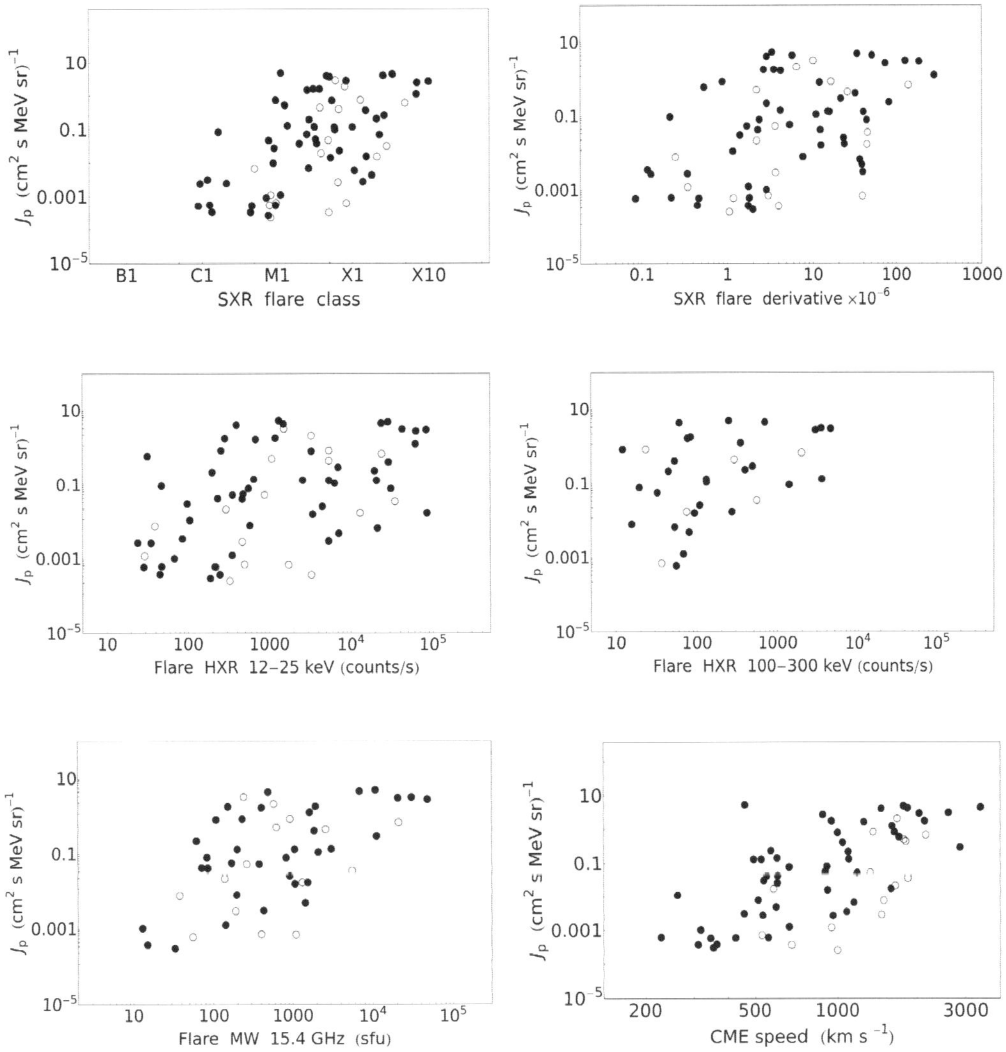

Figure 4: Selected scatter plots of the ~20 MeV peak proton intensity (J_p) and the properties of flare EM emission and/or CME speed. Open circles denote events originating from eastern helio-longitudes, filled circles – western origin events.

4. DISCUSSION

The use of non-thermal flare signatures compared to SXRs when performing correlations with in situ SEP flux is the main objective in this study. Moreover, here we applied different selection criteria compared to Kiplinger (1995), namely we started with a list of in situ proton events identified from ~20 MeV SOHO/ERNE data, performed association with SXR flares and then searched for the HXR signatures. The final number of SFs (reduced to only 70) is primarily due to the observing mode of the HXR instrument used.

HXR emission signatures are a more adequate choice to explore non-thermal particle acceleration processes taking place in SFs compared to thermal SXRs. We used the simplest way (corrected count rates) to quantify the light curves in different energy bands provided by the RHESSI satellite. Lower correlations between the >50 keV HXR count rates and the proton peak intensity as compared to the SXR flare class and CME speed are found. In contrast, when using MW (known to temporarily correlate better with HXRs) and UV emission we obtain results closer to those of SXRs.

A possible reason for the obtained correlation trends could be either a reduced contribution or efficiency drop of the SF acceleration process (in terms of HXR emission) to the in situ proton flux. Alternatively, the sample of 70 events could be biased due to the observational selectivity of the HXR instrument. Both possibilities should be further explored. Moreover, using the peak value of the SXR emission (so-called flare class) could lead to an overestimation of the SF influence to SEP events. In addition to the SF acceleration and emission, the energized particles need to escape the Sun and reach the detector, which could be suppressed, to a degree, in these events due to a specific magnetic field configuration.

An important forthcoming study to be performed is using HXR fluence instead of the peak value in the statistical analysis, as shown by Trottet et al. (2015) that SXR fluence should be used instead of the SXR class. Moreover, HXR flux, contrary to count rates, deduced in terms of model fitting is another step of a future analysis.

Acknowledgements

We acknowledge SOHO/ERNE, RHESSI, GOES, SDO/AIA and RSTN data in our study. SOHO is a project of international collaboration between ESA and NASA. This research is funded by a bilateral collaborative project under agreement between Bulgaria's National Science Fund NTS/AUSTRIA 01/23, 28.02.17 and Austria OeAD Project No. BG 11/2017.

References

Bazilevskaya, G. A.: 2017, *Journal of Physics: Conference Series*, **798 (1)**, 012034.

Benz, A.: 2002, *Plasma Astrophysics. Kinetic Processes in Solar and Stellar Coronae*, second edition. By A. Benz, Institute of Astronomy, ETH Zürich, Switzerland. *Astrophysics and Space Science Library*, **279**, Kluwer Academic Publishers, Dordrecht.

Benz, A. O.: 2017, *Living Reviews in Solar Physics*, **14 (1)**, 2.

Cane, H. V., Richardson, I. G., von Rosenvinge, T. T.: 2010, *Journal of Geophysical Research*, **115 (A8)**, A08101.

Cliver, E. W.: *Central European Astrophysical Bulletin*, **33**, 253–270.

Desai, M., Giacalone, J.: 2016, *Living Reviews in Solar Physics*, **13 (1)**, 3.

Dennis, B. R.; Zarro, D. M.: 1993, *Solar Physics*, **146 (1)**, 177–190

Dierckxsens, M., Tziotziou, K., Dalla, S., Patsou, I., Marsh, M. S., Crosby, N. B., Malandraki, O., Tsiropoula, G.: 2015, *Solar Physics*, **290 (3)**, 841–874.

Fletcher, L.; Dennis, B. R.; Hudson, H. S.; Krucker, S.; Phillips, K.; Veronig, A. et al.: 2011, *Space Science Reviews*, **159 (1/4)**, 19–106.

Guidice, D. A.; Cliver, E. W.; Barron, W. R.; Kahler, S.: 1981, *Bulletin of the American Astronomical Society*, **13**, 553.

Kahler, S. W.: 1982, *Journal of Geophysical Research*, **87**, 3439–3448.

Kiplinger, A. L.: 1995, *The Astrophysical Journal*, **453**, 973–986.

Klein, K.-L., Dalla, S.: 2017, *Space Science Reviews*, **212 (3/4)**, 1107–1136.

Lario, D., Aran, A., Gómez-Herrero, R., Dresing, N., Heber, B., Ho, G. C., Decker, R. B., Roelof, E. C.: 2013, *The Astrophysical Journal*, **767 (1)**, 41.

Lemen, J. R., Title, A. M., Akin, D. J., Boerner, P. F., Chou, C.; Drake, J. F. et al.: 2012, *Solar Physics*, **275 (1-2)**, 17–40.

Lin, R. P., Dennis, B. R., Hurford, G. J., Smith, D. M., Zehnder, A., Harvey, P. R. et al.: 2002, *Solar Physics*, **210 (1)**, 3–32.

Miteva, R.: 2017, Thirteenth International Scientific conference "Space, Ecology, Safety - SES", held 2-4 November 2017 in Sofia, Bulgaria, edited by G. Mardirossian, Ts. Srebrova and G. Jelev, ISSN: 1313-3888, 52–56.

Miteva, R.; Danov, D.: 2017, Ninth Workshop 'Solar Influences on the Magnetosphere, Ionosphere and Atmosphere', proceedings of the conference held 30 May-3 June, 2017 in Sunny Beach, Bulgaria, edited by K. Georgieva, B. Kirov and D. Danov, ISSN 2367-7570, 66–69.

Miteva, R.; Danov, D.: 2018, Tenth Workshop 'Solar Influences on the Magnetosphere, Ionosphere and Atmosphere', proceedings of the conference held 4-8 June, 2018 in Primorsko, Bulgaria, edited by K. Georgieva, B. Kirov and D. Danov, ISSN 2367-7570, 99–104.

Reames, D. V.: 1999, *Space Science Reviews*, **90 (3/4)**, 413–491

Torsti, J., Valtonen, E., Lumme, M., Peltonen, P., Eronen, T., Louhola, M. et al.: 1995, *Solar Physics*, **162 (1/2)**, 505–531.

Trottet, G., Samwel, S., Klein, K.-L., Dudok de Wit, T., Miteva, R.: 2015, *Solar Physics*, **290 (3),** 819–839

Veronig, A. M., Brown, J. C., Dennis, B. R., Schwartz, R. A.; Sui, L., Tolbert, A. K.: 2005, *The Astrophysical Journal*, **621 (1)**, 482–497

Wall, J. V., Jenkins, C. R.: 2003, *Practical statistics for astronomers*, Cambridge observing handbooks for research astronomers, Vol. 3, Cambridge University Press.

Proceedings of the XI Bulgarian-Serbian Astronomical Conference
(XI BSAC) Belogradchik, Bulgaria, May 14-18, 2018
Editors: Milcho K. Tsvetkov, Milan S. Dimitrijević and Momchil Dechev
Publ. Astron. Soc. "Rudjer Bošković" No 18, 2018, 127-134

A SEARCH FOR NEW VARIABLE OBJECTS IN THE FIELD OF OB81 ASSOCIATION IN M31 GALAXY

PETKO NEDIALKOV[1], MITKO CHURALSKI[1,2] and
ANTONIYA VALCHEVA[1]

[1]*Department of Astronomy, Sofia University, 5 J. Bourchier Blvd., 1164 Sofia, Bulgaria*
[2]*Institute of Astronomy and National Astronomical Observatory, Bulgarian Academy of Sciences, 72 Tsarigradsko Chaussee Blvd., 1784 Sofia, Bulgaria*
E-mail: japet@phys.uni-sofia.bg, mvchuralski@yahoo.com,
valcheva@phys.uni-sofia.bg

Abstract. We obtained an astrometric solution with 0.05′′ accuracy for 678 stars in a 5.6′×′5.6′ field within OB81 association in M31 galaxy. Their UB photometry was carried out by Kurtev (2002) on CCD images, obtained with the 2m telescope at NAO-Rozhen. The comparison with the stellar photometry within the same field, published in the LGS M31 catalog (Massey et al. 2016) showed no zero-point difference in B-band, whereas in U-band, a systematic difference of 0.13 ± 0.03 mag has been found.

Thirteen candidates for variable objects have been detected at level, greater than 3σ. Five of them are known variables of the δ Cep type and one star has been detected as IR variable by WISE mission. The remaining 7 candidates are newly detected variables, but only one of them at detection level, greater than 4σ.

1. INTRODUCTION

Almost a century ago, in 1919 the study of variable and transient objects in the M31 galaxy field was initiated by Edwin Hubble who observed with the 2.5m telescope, completed in 1917, at Mount Wilson Observatory. His first results (Hubble, 1925) proved the existence of variables like cepheids and novae in M31 and M33 galaxies. Since then and especially during the last decades that field is a focus of an increasing interest, both amateur's and professional. We are kindly redirecting the reader to the paper 'Time-domain studies of M31' of Lee (2017) which is the latest review of the topic from a professional point of view.

There are three main M31 CCD surveys of point-like sources in the optical, namely McGraw-Hill BVRI (Magnier et. al. 1992), LGS UBVRI (Massey et al. 2006) and HST PHAT (Panchromatic Hubble Andromeda Treasury; Williams et al. 2014). These surveys, equally spaced in time with a step of 10 years, cover a total span of 20 yr. When compared to each other they provide an important although rather sparse information on the variability of much larger time scales than the modern time-domain studies (typically 2-3 yr).

In this pilot search we focus on a small program Rozhen field (see Figure 1) for which we have original CCD photometry. We compared it with the photometry from the LGS by using the magnitude differences to detect variable candidates at detection level, greater than 3σ.

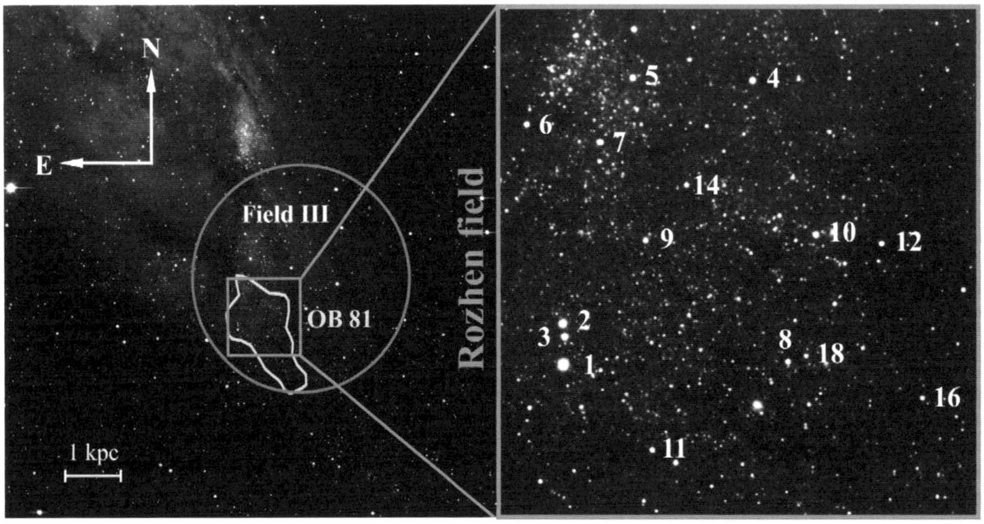

Figure 1: Left: OB 81 association location within LGS (Massey et al. 2006) program field M31F8 together with Field III (Baade and Swope 1965) and program Rozhen field (Kurtev 2002). Right: Zoomed Rozhen field with selected astrometric standards. V-band Mosaic image credit: LGS.

Baade and Swope (1965) discovered 334 variables in their Field III and Rozhen field falls within its boundaries with 75 variables in total, among which 57 are cepheids, 7 eclipsing, 7 semiregular, 4 irregular and peculiar variables. More recently Kodric et al. (2013) confirmed 82 cepheids within Rozhen program field with 61 of them pulsating in the fundamental mode, 6 – in the first overtone and 14 with unidentified pulsation mode and one Population II cepheid.

2. INITIAL DATA SAMPLES

The first source is an unpublished UB stellar photometry and pixel positions by Kurtev (2002) who presented the data only as a color-magnitude diagram graphic. The photometry was carried out on CCD images, obtained with the first professional CCD camera Photometrics mounted on the 2m RCC telescope at NAO-Rozhen. It was kindly provided by the author to our disposal in a machine readable format. Landolt (1992) standards were use to transform the instrumental magnitudes into the Johnson's UB system.

The second source of photometry and astrometry is the revised version of the LGS stellar catalog of M31 galaxy (Massey et al. 2016). It was originally published in 'A Survey of Local Group Galaxies Currently Forming Stars. I. UBVRI Photometry of Stars in M31 and M33' where Massey et al. (2006) presented photometry, carried on the mosaic images of M31, obtained with Mayall 4m telescope at Kitt Peak at US National Observatory, Arizona. The final catalog contains 371 781 entries each of which has been detected at least in the BVR bands together.

We summarize the basic information on the used observation data within Rozhen program field in Table 1.

Table 1: Rozhen field observation data.

Photometry source	Kurtev (2002)	Massey et al. (2016)
Total number of stars	678	5046
Observation dates (B passband)	Aug 6/7 2000	Oct. 4 2000 – Sept. 11 2002
Seeing	1.5′′–1.8′′	0.9′′–1.2′′
Number of stars with UB photometry	678	3946
PSF fitting	DAOPHOT	DAOPHOT
Photometric accuracy at B~21.0 mag	<0.15 mag	<0.015 mag
Total exposure time/per frame	900 s	3000 s
CCD detector size in px	1k×1k	Mosaic 8 × (4k×2k)
Telescope	2m RCC NAO-Rozhen	4m Mayall Kit Peak NAO
FoV per frame	5.6′×5.6′	36′×36′

3. ROZHEN FIELD ASTROMETRIC SOLUTION

The coordinate's cross-identification is substantial for an effective search of variable objects in the Rozhen field. For this purpose, it is necessary to determine the equatorial coordinates of all stars, i.e. to obtain accurate astrometric solution for that particular field. We used 12 bright stars (see Figure 1, right) with known both pixel (Kurtev 2002) and celestial (Massey et al. 2016) coordinates to produce the matched coordinate list (see Table 2) and the standard IRAF routine *ccmap* in order to calculate the image field solution.

Later on we used *cctran* IRAF routine to transform all 678 stars pixel coordinates from the photometry of Kurtev (2002) to celestial coordinates on the base of the *ccmap* Rozhen field solution.

The accuracy of the computed equatorial coordinates of 678 stars is ~0.05′′ both on R.A. and DEC. It corresponds to a radius of ~0.07′′. A small rotation angle of the CCD detector with respect to the North pole is estimated to 1.5°.

Table 2: Matched Rozhen/LGS coordinate list used to compute the Rozhen field astrometric solution.

Star number	X [pixels]	Y [pixels]	RA h m s	DEC ° ′ ″
3	137.9	372.9	00:40:34.89	40:31:35.1
4	581.7	918.7	00:40:23.22	40:34:28.1
5	313.5	933.6	00:40:30.53	40:34:30.5
6	68.5	840.5	00:40:37.14	40:33:59.5
7	234.3	796.1	00:40:32.58	40:33:47.2
8	639.4	300.7	00:40:21.20	40:31:17.0
9	329.1	578.1	00:40:29.85	40:32:40.4
10	713.8	576.5	00:40:19.37	40:32:43.1
12	861.7	552.2	00:40:15.35	40:32:36.9
14	425.7	695.3	00:40:27.31	40:33:17.6
16	940.6	210.3	00:40:12.94	40:30:51.3
18	681.4	311.9	00:40:20.06	40:31:20.8

4. PHOTOMETRIC SYSTEMS COMPARISION

We adopted value of 0.21′′=3×0.07′′ as a 3σ clipping search radius and successfully cross-identified 387 stars from Kurtev (2002) with a single counterpart in LGS.

The direct comparison of B-band magnitudes where the consistency is good is shown in Figure 2 and for U-band where a systematic shift exists – on Figure 3. As expected, the dispersion of the stellar magnitudes increases with the magnitude itself. That increase, however, is mainly due to the larger errors of the Rozhen's photometry. The B-band magnitude histogram indicates a completeness magnitude of that photometry B=21.5 mag where Kurtev (2002) pretends for error of ~0.15 mag. At the same B-magnitude the photometry of LGS photometry (which goes much deeper) is only ~0.025 mag. An approximation of the difference (B_{lgs}– B_r) within the magnitude range 20.8 mag < B_{lgs} < 21.2 mag with a Gaussian function shows (see Figure 4) mean value of –0.03 ± 0.02 mag and a standard deviation of 0.20 mag proving that accuracy of the Rozhen photometry is overestimated. Analogous approximation at other magnitude ranges with taking outliers into account allows a precise quantitative analysis of the budget errors of Kurtev's photometry, which is not known a priori.

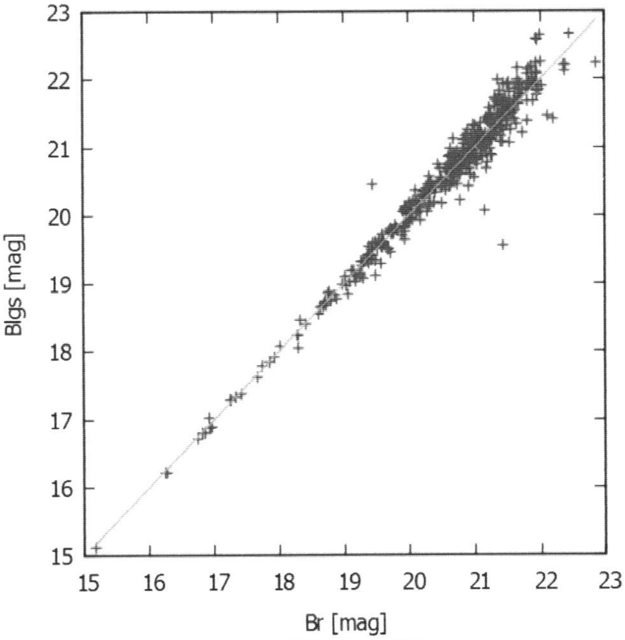

Figure 2: B-band comparison between photometry of Kurtev (2002) and that of Massey et al. (2016) for 387 stars in common.

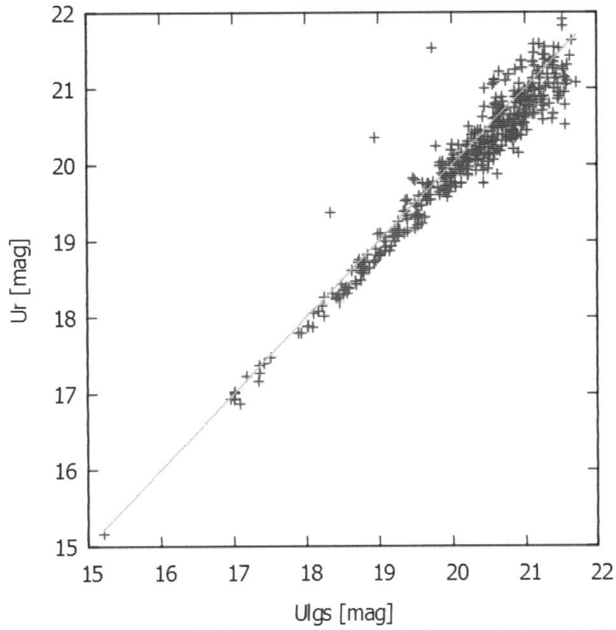

Figure 3: U-band comparison between photometry of Kurtev (2002) and that of Massey et al. (2016) for 387 stars in common.

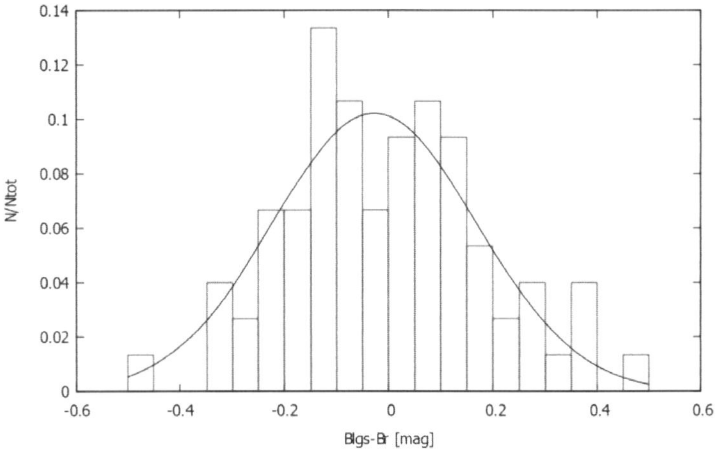

Figure 4: Approximation of the difference (B_{lgs}– B_r) histogram within the magnitude range 20.8 mag < B_{lgs} < 21.2 mag with a Gaussian function. The mean value is –0.03 ± 0.02 mag and the standard deviation – 0.20 ± 0.02 mag.

When two photometries are compared it is important not only to check the magnitude differences for a systematic shift but also to see whether these differences vary with color. We performed such tests and found that there is no strong correlation of either B or U differences with color. We also found that there is no significant shift in B-band, whereas in U-band, the systematic difference of ΔU = 0.13 ± 0.03 mag is three times larger than the zero-point magnitude accuracy of 0.04 mag claimed by Kurtev (2002).

5. VARIABLE OBJECTS DETECTION

The detection of variable objects is based on the B-band magnitudes difference modulus $|B_{lgs}–B_r|$, normalized to the error of that difference:

$$\sigma = \sqrt{\sigma_{Blgs}^2 + \sigma_{Br}^2}.$$

We call that quantity 'level of detection'. The Rozhen field photometry B_r is taken from Kurtev (2002) and LGS photometry B_{lgs} and its error σ_{Blgs}– from the revised version of the M31 catalog (Massey et al. 2016). The typical error of the Rozhen magnitudes σ_{Br} at a given B_r magnitude is derived from the standard deviation of maginitudes difference (B_{lgs}–B_r), described in the previous section.

We plotted on Figure 5 the B-band magnitudes difference modulus $|B_{lgs}–B_r|$ as function of B_{lgs} for all successfully cross-identified 387 stars from Kurtev (2002) with a single counterpart in LGS. We also constructed approximating polynomials for the quantities: σ (at two levels of detection: 1σ и 3σ), σ_{Blgs} and σ_{Br} as a function B_{lgs} magnitude. As clearly seen in the same figure, the error of the magnitudes difference σ is dominated by the error of the Rozhen photometry σ_{Br}. We found 13 candidate variable objects at the detection level, greater than 3σ and list them in Table 3.

Table 3: Variable objects in the program Rozhen field detected at level, greater than 3×standard deviation of the magnitude differences (B_{lgs}– B_r).

№	*RA [deg]*	*DEC [deg]*	*σ*	*Name*	*Ref.*	*Var. Type*
1	10.09802	40.51178	3.2	J004023.54+403042.4	LGGS	
2	10.13827	40.51916	4.5	J004033.12+403107.7	LGGS	
3	10.14069	40.51204	3.7	M31 V0418	GCVS	CEP
4	10.08728	40.51318	3.1	J004020.96+403047.4	LGGS	
5	10.10084	40.53334	3.3	J004024.23+403200.2	LGGS	
6	10.08655	40.55610	3.0	M31 V0285	GCVS	DCEP
7	10.08293	40.51323	3.6	J004019.93+403047.7	LGGS	
8	10.08832	40.52137	3.7	J004021.21+403117.1	LGGS	
9	10.14698	40.55776	9.4	M31 V0438 J010.1469+40.5577	GCVS Kod13	DCEP FM
10	10.13570	40.51632	12.9	M31 V0411	GCVS	DCEP
11	10.09008	40.54896	3.1	J004021.64+403256.5	LGGS	
12	10.04950	40.50658	6.1	M31 V0215 J010.0495+40.5065	GCVS Kod13	DCEP FM
13	10.11720	40.50244	6.0	J004028.12+403008.5	WISE	high prob. var

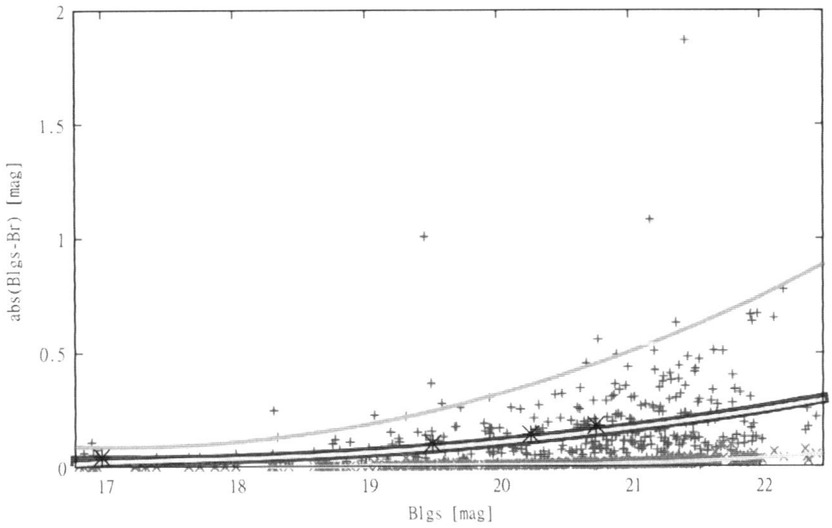

Figure 5: B-band magnitudes difference modulus $|B_{lgs}–B_r|$ as function of B_{lgs} for all successfully cross-identified 387 stars (red crosses) from Kurtev (2002) with a single counterpart in LGS (Massey et al. 2016). Four approximating functions of B_{lgs} of are also plotted: (B_{lgs}–B_r) standard deviation (blue line); (B_{lgs}–B_r) 3×standard deviation (green line), B_{lgs} standard deviation (azure line through the pink squares) and B_r standard deviation (yellow line). All 13 detected variable objects lie above the solid green line, representing 3σ detection level.

Five of them are known variables of the δ Cep type and one star has been detected as IR variable by WISE mission. The remaining 7 candidates are newly detected variables, but only LGSJ004033.12+403107.7 at a detection level, greater than 4σ. Our test recovers only ~6% of all known variables within Rozhen field which is not unexpected since we compare only two epoch magnitudes in a single pass-band.

Rozhen field was observed on Aug 6/7 2000 while the time span of the LGS B-band observations of the same field covers three consecutive years (Oct. 4 2000 – Sept. 10 2002) for the program fields F7, F8 and F9. Thus, most of the LGS B-magnitudes are averaged upon several independent but indistinguishable measurements, a problem that will be solved in a forthcoming paper. There we will exploit magnitude estimates related to unique epochs of observations rather than averaged over a number of detections.

5. CONCLUSIONS

The intention of this pilot search was to show the potential of two epoch comparison of photometry for finding new variable objects in the field of OB81 – a huge stellar complex in the south-eastern part of M31 galaxy disk. We used UB photometry and pixel coordinates of Kurtev (2002) and UB photometry and celestial coordinates from LGS (Massey et al. 2016) to obtain an astrometric solution of the Rozhen field and to check the magnitude differences of coinciding objects. Significant systematic shift of 0.13 ± 0.03 mag has been found in U-band.

Thirteen candidates for variable objects have been detected at level, greater than 3σ. Five of them are known variables of the δ Cep type and one star has been detected as IR variable by WISE mission. The remaining 7 candidates are newly detected variables, but only one of them, namely LGSJ004033.12+403107.7, at detection level, greater than 4σ.

Acknowledgments

The authors are thankful to Dr. Radostin Kurtev for the opportunity to use the unpublished stellar photometric and positional data for the area of the stellar complex OB81.

This paper has been accomplished with the financial support from the grant 80-10-67/2018 with the Science Fund of the Sofia University and the grant DN18/10 with the National Science Fund of Bulgaria.

References

Baade, W., Swope, H. H.: 1965, *Astron. J.*, **70**, 212–268.
Hubble, E. P.: 1925, *The Observatory*, **48**, 139–142.
Kodric, M., Riffeser, A., Hopp, U. et al.: 2013, *Astron. J.*, **145**, 106–127.
Kurtev, R. G., 2002, *Publications of the Astron. Observatory of Belgrade*, **73**, 163–168.
Lee, C.-H.: 2017, *Astron. Rev.*, **12**, 1–23
Magnier, E. , Lewin, W., van Paradijs, J. et al.: 1992, *Astron. Astrophys.*, **96**, 379–388.
Massey, P., Olsen, K., Hodge, P. et al.: 2006, *Astron. J.*, **131**, 2478–2496.
Massey, P., Neugent, K., Smart, B.: 2016, *Astron. J.,* **152**, 62–78.
Williams, B., Lang, D., Dalcanton, J. et al., 2014, *Astron. J.,* **215**, 9-43.

**Proceedings of the XI Bulgarian-Serbian Astronomical Conference
(XI BSAC) Belogradchik, Bulgaria, May 14-18, 2018
Editors: Milcho K. Tsvetkov, Milan S. Dimitrijević and Momchil Dechev
Publ. Astron. Soc. "Rudjer Bošković" No 18, 2018, 135-138**

CHEMI-IONIZATION/RECOMBINATION ATOMIC PROCESSES IN THE AGNs BROAD-LINE REGION

VLADIMIR A. SREĆKOVIĆ[1], MILAN S. DIMITRIJEVIĆ[2]
and LJUBINKO M. IGNJATOVIĆ[1]

[1]*Institute of Physics, University of Belgrade, P.O. Box 57, 11001 Belgrade, Serbia*
[2]*Astronomical Observatory, Volgina 7, 11060 Belgrade, Serbia*
[3]*LERMA, Observatoire de Paris, PSL Research University, CNRS, Sorbonne Universités, UPMC (Univ. Pierre \& Marie Curie) Paris 06, 5 Place Jules Janssen, 92190 Meudon, France*
E-mail: vlada@ipb.ac.rs, mdimitrijevic@aob.rs, vlada@ipb.ac.rs

Abstract. The chemi-ionization and chemi-recombination processes in atom - Rydberg atom collisions, are considered as factors of influence on the atom excited-state populations and ionization level in clouds in Broad Line Region (BLR) of active galactic nuclei (AGN). The presented results are related to the moderately ionized layers of dense parts of the BLR clouds. The preliminary analyses show that the considered chemi-ionization/recombination processes should have a very significant influence on the optical properties of some regions in AGN.

1. INTRODUCTION

Elementary processes in plasma phenomena traditionally attract astrophysicist's attention (see e.g. Barklem 2007). For diagnostic methods needed in order to estimate physical conditions in the broad line region (BLR) in active galactic nuclei (AGN) the study of the influence of various atomic and molecular processes may be of interest (Netzer 1990). Consequently, it is of interest to investigate the influence of chemi-ionization/recombination processes in similar conditions that probably hold in some parts of BLR clouds in AGN.

When an electron is excited into a high lying Rydberg state, with large principal quantum number, even inelastic thermal collisions can be sufficiently energetic to lead to ionization reactions (see Mihajlov et al. (2011, 2012)). These types of reactions can be classified as chemi-ionization reactions (associative ionization and non-associative ionization).

$$H^{*}(n)+H \Rightarrow e+H_{2}^{+} \qquad (1)$$

$$H^{*}(n)+H \Rightarrow e+H^{+}+H \qquad (2)$$

and the corresponding inverse recombination processes:

$$H_{2}^{+}+e \Rightarrow H^{*}(n)+H \qquad (3)$$

$$H^{+}+H+e \Rightarrow H^{*}(n)+H \qquad (4)$$

where H and H^+ are atoms and atomic ions in their ground states, $H^*(n)$ is atom in a highly excited (Rydberg) state with the principal quantum number $n \geq 2$.

Using these partial rate coefficients, we will determine the total ones, namely,

$$K_{ci}^{(ab)}(n,T)=K_{ci}^{(a)}(n,T)+K_{ci}^{(b)}(n,T). \qquad (5)$$

which characterizes the efficiency of the considered chemi-ionization and chemi-recombination processes together (for details see paper of Mihajlov et al. (2015)). These rate coefficients enable inclusion of these excited-state reactions in modeling AGNs. Using rate coefficients, we can obtain the total chemi-ionization and chemi-recombination fluxes caused by the processes (1) and (2) i.e.,

$I_{ci}(n,T)=K_{ci}(n,T)\cdot N_{n}N_{1}$ and $I_{cr}(n,T)=K_{cr}(n,T)\cdot N_{1}N_{i}N_{e}$. (6)

Using these expressions, we can calculate quantities $Fi;ea(n;T) = Ici(n;T)/Ii;ea(n;T)$ which characterize the relative efficiency of partial chemi-ionization processes (1) and (2) together and the impact electron-atom ionization Ii;ea as one of most important concurrent process in the considered plasma, Mihajlov et al. (2011).

2. RESULTS AND DISCUSSION

The values of the calculated total chemi-ionization and recombination rate coefficients $K_{ci}(n;T)$ and $K_{cr}(n;T)$ are illustrated by Fig. 1. The results cover the regions of quantum numbers and temperatures which are relevant for considered cases. This enabled the correct inclusion of these processes in the modeling of AGNs since in the existing literature processes (1) and (2) are poorly defined and described.

In order to enable the better and more adequate use of data, we have derived for the rate coefficients a simple and accurate fitting formula based on a least-square method, which is logarithmic and represented by a second-degree polynomial:

$$\log(K_{ci}(T))=a_{1}+a_{2}\log(T)+a_{3}(\log(T))^{2} \qquad (7)$$

The fits are valid over the temperature range of 4000 K $\leq$T $\leq$ 20000 K and *n* up to 20. Also, it is possible that the fit is applicable outside this area, but with caution. In the Tab. 1 are presented the fits for $2 \leq n \leq 14$.

The preliminary analyses of the behavior of the quantities *F*i;ea(*n;T*) as functions of temperature were done. It follows that the efficiency of the considered chemi-ionization processes in comparison with the electron-atom impact ionization increases with increasing temperature and could be important for *n* up to 10. We need to continue research in the direction of higher values of *n* and region of higher temperatures.

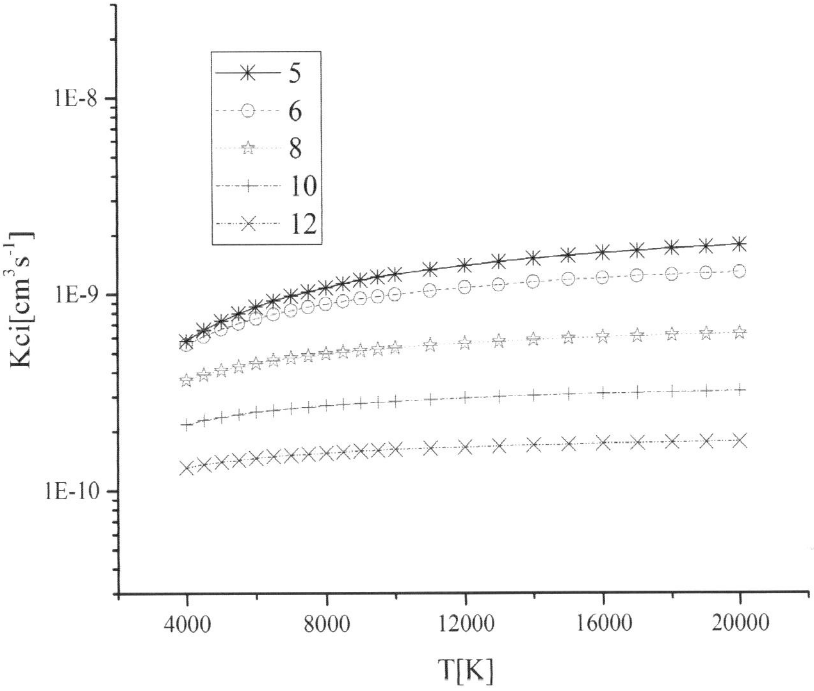

Figure 1: Values of collisional ionisation rate coefficients Kci[cm3/s] (2.1) as a function of n and T.

3. CONCLUSIONS

The obtained results about chemi ionization/ recombiontion processes could be used for determination of limiting high densities in clouds in AGN BLR and NRL regions and for the improvement of modelling of dense moderately ionized layers in them. Of course, the presented values of the corresponding rate coefficients are also very useful for the modelling and analysis of similar layers in the photospheres of Sun and solar like stars.

Table 1: The fits Eq.(7) of the rate coefficient Eq.(5). A portion is shown here for guidance regarding its form.

n	a1	a2	a3
2	-56.410500	19.716100	-2.0390600
3	7.0794400	-8.930940	1.2185400
4	12.4053000	-11.40770	1.5155100
5	-19.6284000	4.7246900	-0.5105980
6	-17.1562000	3.5990500	-0.3900280
7	-15.6337000	2.8749300	-0.3124150
8	-14.5801000	2.3479700	-0.2552700
9	-13.8527000	1.9646600	-0.2135550
10	-13.3764000	1.6961900	-0.1845360
11	-13.0238000	1.4848800	-0.1616660
12	-12.6763000	1.2725000	-0.1378230
13	-12.5340000	1.1620000	-0.1261860
14	-12.4537000	1.0833400	-0.1182720

Acknowledgments

This work is partially supported by Ministry of Education, Science and Technological Development of the Republic of Serbia under the grants III 44002, and 176002.

References

Barklem P. S.: 2007, „Non-LTE Balmer line formation in late-type spectra: effects of atomic processes involving hydrogen atoms“, *A&A*, **466**, 327.

Mihajlov A. A., Ignjatović L.M., Srećković V. A., Dimitrijević M. S.: 2011, „Chemi-ionization in solar photosphere: influence on the hydrogen atom excited states population“, *ApJS*, **193**, 2.

Mihajlov A. A., Srećković V. A., Ignjatović L.M., Klyucharev, A. N., Sakan, N. M.: 2015, „Non-Elastic Processes in Atom Rydberg-Atom Collisions: Review of State of Art and Problems“, *J. Astrophys. Astron.*, **36**, 623.

Mihajlov A. A., Srećković V. A., Ignjatović L. M., & Klyucharev A. N.: 2012, „The chemi-ionization processes in slow collisions of Rydberg atoms with ground state atoms: mechanism and applications“, *J. Clust. Sci.*, **23**, 47.

Netzer H.: 1990, *AGN emission lines in Active Galactic Nuclei*, pp. 57–158. Springer.

Proceedings of the XI Bulgarian-Serbian Astronomical Conference
(XI BSAC) Belogradchik, Bulgaria, May 14-18, 2018
Editors: Milcho K. Tsvetkov, Milan S. Dimitrijević and Momchil Dechev
Publ. Astron. Soc. "Rudjer Bošković" No 18, 2018, 139-146

A NEW POTENTIAL OF MILKY WAY GIVEN ANALYTICALLY

MILAN STOJANOVIĆ[1], SLOBODAN NINKOVIĆ[1],
NEMANJA MARTINOVIĆ[1], MILJANA D. JOVANOVIĆ[1]
and GABRIJELA MARKOVIĆ[2]

[1]*Astronomical Observatory, Volgina 7, 11060 Belgrade, Serbia*
[2]*Faculty of Mathematics, University of Belgrade, Studentski trg 16,*
11000 Belgrade, Serbia
E-mail: mstojanovic@aob.rs

Abstract. A mass distribution model for the Milky Way is presented. The potential of this galaxy is contributed by four subsystems: bulge, innermost dark matter, disc and outer dark matter. For each of them the potential is given analytically by using elementary functions. The values of the model parameters are specified by fitting an assumed rotation curve. The model is foreseen to be used in the determination of galactocentric orbits.

1. INTRODUCTION

The Galaxy structure is difficult to discern since we are observing from within. Traditionally, it has three components bulge, probably with bar, disc and dark halo (Dehnen & Binney, 1998). Each component contributes to gravitational field. A new idea was introduced by Iocco et al. (2015) where they showed that current observational data strongly disfavour baryons as the sole contribution to the Galactic mass budget, even inside the solar circle. In this paper we will show new potential of Milky Way consisting of four components, where innermost dark matter is the newest addition, together with new potentials for some components.

2. COMPONENTS OF THE MILAKY WAY

In order to construct a gravitational potential of the Milky Way analyti-cally, we restrict ourselves to axisymmetric models of components of the Milky Way and steady state. Also we use only elementary functions for describing the potential. As for the number of components, we use four in this study. Three main components being bulge, disc and dark corona, as in many previous studies, and the fourth one innermost dark matter. We use completely new formulas for disc and dark corona, that were recently published (Ninković, 2015; Ninković, 2017).

2.1. Innermost dark matter and bulge

For innermost dark matter (IDM) we use potentials of some simple systems, nevertheless when combined with other components of the Milky way, we get good results on final rotation curve, especially since it is known that many previous models when compared to measurements from the inner few kpc of Galaxy do a poor job of reproducing them. The potential of IDM inside limiting radius r_{IDM} is modelled as homogenous sphere with constant density ρ and mass $M_{IDM} = \frac{4}{3}\pi\rho r_{IDM}^3$, so we have that innermost dark matter contribution to circular speed (in $km^2 s^{-2}$) is:

$$u_{cIDM} = \sqrt{\frac{GM_{IDM}}{r_{IDM}^3}} R, R\; r_{IDM},$$

(1)

where *G* is the universal gravitation constant and R is variable, distance to the axis of symmetry (*z*). Outside the homogenous sphere we use Keplerian or point mass potential:

$$u_{cIDM} = \sqrt{\frac{GM}{R}}, R > r_{IDM}.$$

(2)

In case of a spherically symmetric subsystem, such as bulge (B), the generalised isochrones potential formula is often used that was proposed in 1973 (Kuzmin & Veltmann, 1973):

$$\Pi_B = \frac{GM_B}{r_a + \sqrt{r_b^2 + r^2}}.$$

(3)

Here Π is potential, M_B is total mass of bulge, *r* is the distance to the Galactic centre and r_a and r_b are constants with dimension of length. Finally, we have bulge contribution to circular speed:

$$u_{cB} = \sqrt{\frac{GM_B R^2}{\sqrt{r_b^2 + R^2}\left(r_a + \sqrt{r_b^2 + R^2}\right)^2}}.$$

(4)

2.2. Disc

The potential of the flattened stellar subsystem, especially disc (D) is often represented by Miyamoto-Nagai (1975) formula:

$$\Pi = \frac{GM_D}{R_{MN}},$$

(5)

$$R_{MN} = \sqrt{R^2 + \left(a + \sqrt{z^2 + b^2}\right)^2}.$$

Here potential is a function of two arguments R and z, while a and b are constants. In recent study done by Ninković (2015) a new potential formula applicable to flattened system is presented. It is a modification of the Miyamoto-Nagai potential and can easily be applied to exponential discs. It consists of a new term R_N, which is also function of the same variables, added in denominator of eq. (5).

$$\Pi_D = \frac{GM_D}{R_{MN} - R_N}.$$

(6)

This new term introduce new constants, disc scale length R_d and dimensionless c_1, c_2 and γ_1, γ_2.

$$R_N = \frac{1}{2} R_d \left[\left(1 + \frac{R^2}{c_1^2}\right)^{\gamma_1} + \left(1 + \frac{z^2}{c_2^2}\right)^{\gamma_2}\right].$$

(7)

For purpose of creating new four-component potential of Milky Way we adopted this formula for disc. Then disc contribution to circular speed is:

$$u_{cD} = \sqrt{\frac{GM_D R}{\left[\sqrt{R^2 + (a+b)^2} - \frac{1}{2} R_d \left(1 + \left(1 + \frac{R^2}{R_d^2}\right)^{\gamma_1}\right)\right]^2}}.$$

(8)

$$\cdot \sqrt{\frac{r}{\sqrt{R^2 + (a+b)^2}} - \gamma_1 \frac{R}{R_d}\left(1 + \frac{R^2}{R_d^2}\right)^{\gamma_1 - 1}}.$$

2.3. Dark corona

For density of dark corona (DC) for this model, we use formula (5) from paper (Ninković, 2017) which was the first time such formula was published, and its reproduced here:

$$\rho_{DC} = \rho_0 \left(\frac{1}{1 + \xi^k} - \frac{1}{1 + \xi_l^k}\right).$$

(9)

Case $k = 3$ is analysed in details in original paper and we chose this as best option for our model. From (9) it follows that cumulative mass for DC is:

$$M_{DC} = 4\pi\rho_0 r_c^3 \left[\frac{ln(1+\xi^3)}{3} - \frac{\xi^3}{3(1+\xi_l^3)}\right].$$
(10)

In eq. (9) k is supposed to be a natural number and for practical reasons $2 \leq k \leq 4$. In eq. (9) and (10) r_c is corona scale length, r_l is corona limiting radius, $\xi = \frac{r}{r_c}$, $\xi_l = \frac{r_l}{r_c}$. As seen from eq.(9), at $\xi = \xi_l$ we get $\rho = 0$, and if $\xi > \xi_l$ then $\rho = 0$. Thus M_{DC} is equal to total mass for $\xi = \xi_l$. Now we can obtain the potential for DC:

$$\Pi = \frac{GM_{DC}}{R} + 4\pi G\rho_0 r_c^2 \left\{\frac{1}{6} ln \frac{\frac{4}{3}\left(\xi_l - \frac{1}{2}\right)^2 + 1}{\frac{4}{3}\left(\xi - \frac{1}{2}\right)^2 + 1} + \frac{\sqrt{3}}{3} arctg\left[\frac{2\sqrt{3}}{3}\left(\xi_l - \frac{1}{2}\right)\right] - \frac{\sqrt{3}}{3} arctg\left[\frac{2\sqrt{3}}{3}\left(\xi - \frac{1}{2}\right)\right] - \frac{1}{3} ln \frac{\xi_l + 1}{\xi + 1} - \frac{\xi_l^2 - \xi^2}{2(1+\xi_l^3)}\right\}. (11)$$

In view of eq. (10) the circular speed of the dark corona is:

$$u_{cDC} = \sqrt{\frac{GM_{DC}}{R}}.$$
(12)

3. RESULTS

In order to obtain values for all free parameters in our models of components of Milky Way we compare our results with up-to-date compilation of Milky Way rotation curve measurements presented in the paper (Iocco et al, 2015). The data points on the Figure 1. are taken from mentioned paper, while three lines overplotted on the graph are our best fit models of the rotation curve measurements. Each model, with all four components shown, is also presented separately on Figure 2. These models are obtained using four-component analytical model discussed in previous section. This is first time such model is used and we show that fourth component improves our fit especially in inner part of the Milky Way where all previous models fail to successfully fit observational data. All parameters for these three models are presented in Table 1.

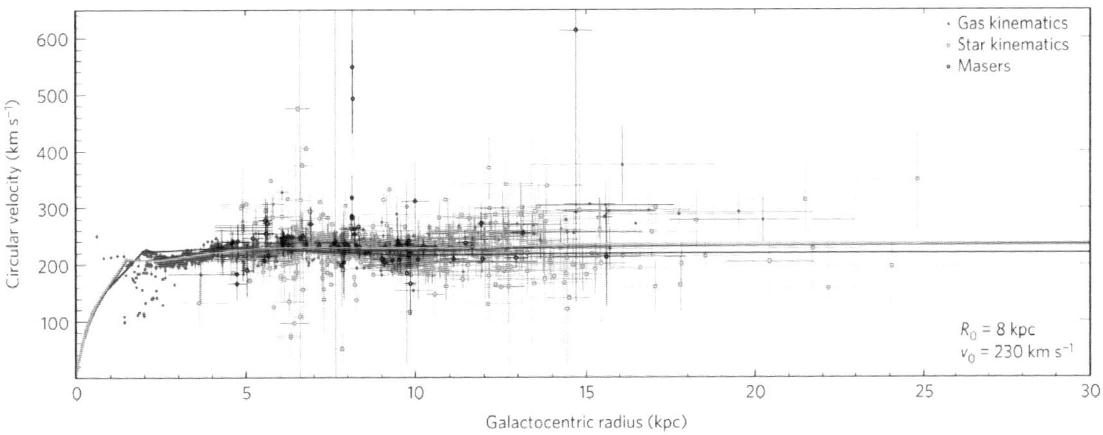

Figure 1: Data points are taken from (Iocco et al, 2015) while three overplotted lines are our best fit models of Milky Way rotation curve. Parameters for these models are given in Table 1.

We have discussed two variants concerning the values for $R_\odot$ and $u_c(R_\odot)$, mentioned very often in the literature: $R_\odot = 8.5$ kpc, $u_c(R_\odot) = 220$ km/s and $R_\odot = 8.0$ kpc, $u_c(R_\odot) = 230$ km/s where the latter one corresponds to the values given by Iocco et al. (2015) for comparison.

In the case of innermost dark matter region, we found that the best fit for inner part of Milky Way rotation curve is produces for $7 < M_{IDM} < 10$ billion solar masses and limiting radius for homogenous sphere $1.5 \leq r_{IDM} \leq 2$ kpc.

Table 1: Parameters that were used in all models. Units: Total mass of each component $M \times 10^9 M_\odot$; ρ_0 in $M_\odot pc^{-3}$; $r_{IDM}, r_a, R_d, R_\odot$ in kpc; $u_c(R_\odot)$ in km/s.

		Model 1 Blue line	Model 2 Green line	Model 3 Red line
$R_\odot$ $u_c(R_\odot)$		8.5 220	8.0 230	8.0 227
Innermost dark matter (IDM)	M_{IDM} r_{IDM}	9.5 2	8 1.5	7.0 1.5
Bulge (B)	M_B r_a	10 0.5	8.5 0.5	7.5 0.5
Disc (D)	M_D R_d γ_1	58 2.85 -0.1	69 3.2 -0.05	66 2.9 -0.1
Dark corona (DC)	ρ_0	0.006	0.007	0.007

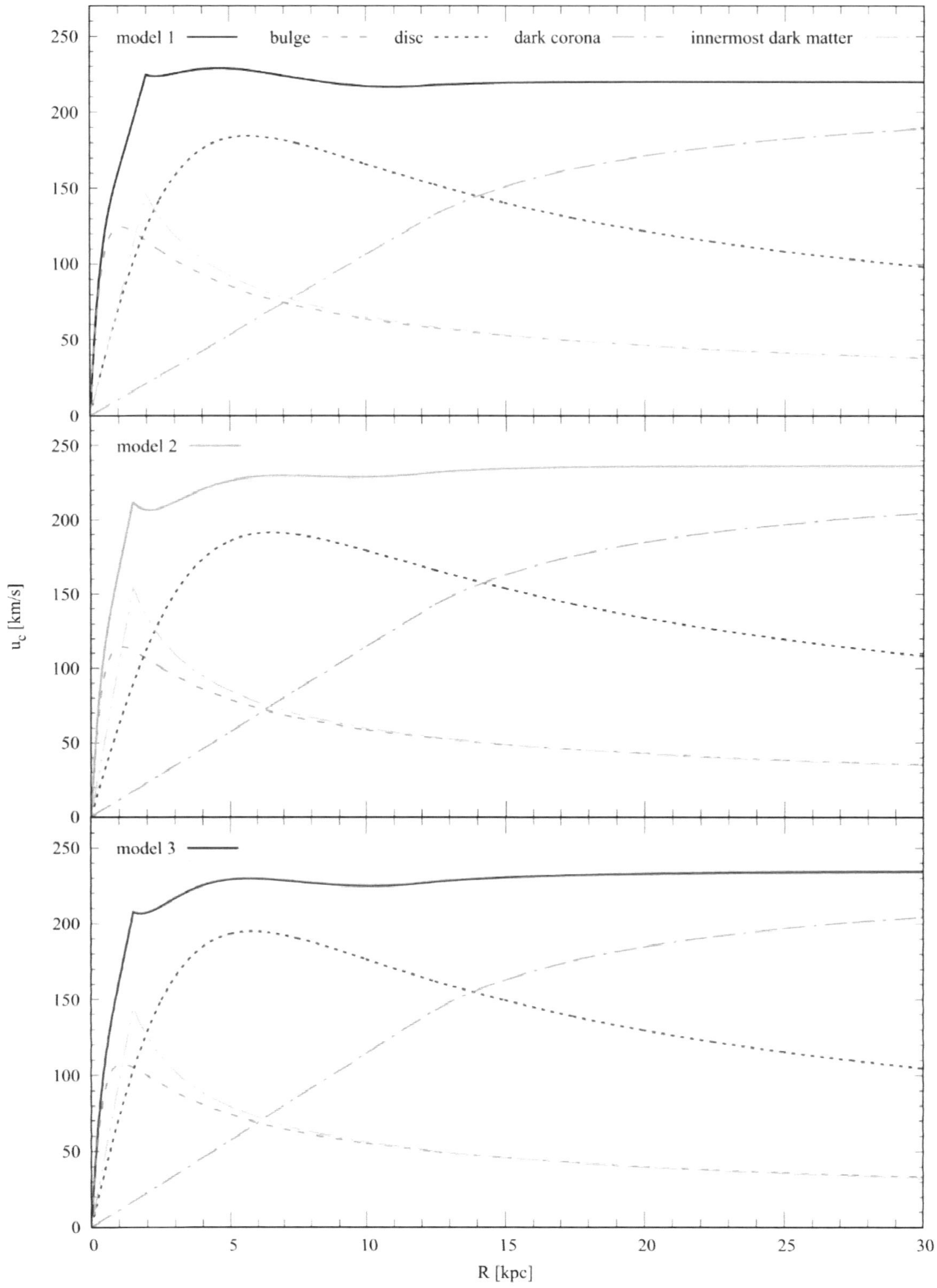

Figure 2: Rotation curve for models presented with contribution from each of four components separately.

For bulge total mass we used a few different values starting from 7.5 up to 10 billion solar masses. In this case we used that $r_a = r_b$ (eq. (3)) and using accordingly formulae for total mass and density, it is easily obtained that within r_a, about 12% of the total mass is contained and that the ratio of the density at the centre to that at r_a is about 3.23.

When it comes to Galactic disc, starting value is surface density $\sigma(R_\odot)$ where $R_\odot$ is galactocentric distance of the Sun. For surface density, we take a value well known in literature $\sigma(R_\odot) = 50M_\odot$ pc^{-2}. In eq. (7) it is also acceptable to substitute c_1 with R_d. In fitting the rotation curve we also determined that best value for ratio $\frac{a+b}{R_d}$ is 2.1. Since this component is flattened, it must satisfy $a \gg b$. For disc scale length we tested interval $2.5 < R_d \leq 3.5$ Exponent γ_1 cannot exceed 0.5 and should be small negative number. Total mass of disc M_D is varied between 50 and 70 billion solar masses.

For dark corona, one should first decide on parameter k, since it should be in interval $[2,4]$. If $k = 3$ is chosen then total mass will depend only on three parameters: ρ_0, r_c and r_l. We find this to be best option when dark corona is modelled. Formula (11) is applied to the circle outside $R = R_\odot$ and we assumed fixed values for $r_c = 2R_\odot$ kpc ($\xi = 0.5$) and $\xi_l = 5.665$.

Limiting case for total mass of Milky Way galaxy was constrained by recent finds of Patel et al. (2018). They showed, by using the Bayesian framework to include all Milky Way satellites with measured 6D phase space information and applying it with the Illustris-Dark simulation, that the Galaxy's total mass should be constrained to $0.85^{+0.23}_{-0.26} \times 10^{12} M_\odot$.

4. CONCLUSION

When Milky Way mass models are discussed most often there are three main contributors to the potential: the bulge, the disc and dark corona. Here we presented a new model with forth contributor innermost dark matter. We show that this could be a missing part in order to better fit rotation curve of Milky Way to most recent observational data, especially in the inner part of Milky Way. For all of them steady state has been assumed and potentials are all given analytically by using only elementary functions. This gives big advantage in making numerical integrator for calculating galactocentric orbits of stars. Such integrators work fast and can handle huge databases which is important in light of new results given by GAIA mission.

Innermost dark matter region is a new component that we introduced here so we used simplest model. The potential of IDM is represented by homogenous sphere. Bulge and disc are observable components of Milky Way and as such these components are best analysed and potential of these components are most complicated. Bulge is represented by generalised isochrones potential formula which is often used, while for the disc we used completely new exponential formula given by Ninković. It is modification of well-known Miyamoto-Nagai formula. Modelling of dark corona is harder since only constraints due to seen matter can be used. Here we use just one case of more complicated new formula given by Ninković recently. This formula should satisfy requirement that there is a prominent maximum near the centre due to the bulge model, but also provides high values for the circular speed in outer parts due to the dark matter model being accepted here.

Acknowledgments

This research has been supported by the Serbian Ministry of Education, Science and Technological Development (Project No 176011 "Dynamics and Kinematics of Celestial Bodies and Systems" and Project No 176021 "Visible and Invisible Matter in Nearby Galaxies: Theory and Observations").

References

Dehnen, W., Binney, J.: 1998, *MNRAS*, **294**, 429-438.
Hernquist, L.: 1990, *ApJ*, **356**, 359-364.
Iocco, F., Pato, M., Bertone, G.: 2015, *Nature Physics*, **11**, 245.
Kuzmin, G. G., Veltmann, Ü. I. K.: 1973, *Publ. Tartuskoj astrof. Observatorii*, **40**, 281-323.
Miyamoto, M., Nagai, R.: 1975, *PASJ*, **27**, 533.
Ninković, S.: 2015, *Publications of the Astronomical Society of Australia (PASA)*, **32**, e032.
Ninković, S.: 2017, *Open Astronomy*, **26 (1)**, 1.
Patel, E. et al.: 2018, *American Astronomical Society*, **232**, 402.03.

Proceedings of the XI Bulgarian-Serbian Astronomical Conference (XI BSAC) Belogradchik, Bulgaria, May 14-18, 2018
Editors: Milcho K. Tsvetkov, Milan S. Dimitrijević and Momchil Dechev
Publ. Astron. Soc. "Rudjer Bošković" No 18, 2018, 147-157

EVOLUTION OF THE ACCRETION STRUCTURE IN THE SYMBIOTIC BINARY BF CYGNI DURING ITS LAST OPTICAL OUTBURST BEGAN IN 2006

NIKOLAI A. TOMOV and MIMA T. TOMOVA

Institute of Astronomy and National Astronomical Observatory,
Bulgarian Academy of Sciences,
BG-4700 Smolyan, POBox 136
E-mail: tomov@astro.bas.bg, mtomova@astro.bas.bg

Abstract: The last optical eruption of the symbiotic eclipsing binary BF Cyg began in 2006 and continues up to the present time. In 2006 its brightness rose sharply reaching a maximum and after that gradually decreased reaching a minimum in 2014. After that it rose at two times again - in the beginning of 2015 and 2017. Its orbital photometric minima determined by eclipses were seen during this eruption. The evolution of the accretion structure surrounding the compact object and changing the depth of the orbital minimum is investigated on the basis of a broad band $UBVR_CI_C$ photometric data and high resolution spectra in the range of the Balmer line Hα. The parameters of this structure are determined at the times of the photometric orbital minima.

1. INTRODUCTION

The symbiotic system BF Cyg is an eclipsing binary and the eclipses are observed in its both states the quiescent and active ones. Its photometric orbital minimum is due to both an eclipse of the compact companion and an occultation of the circumbinary nebula (Tomov et al. 2015, hereafter Paper I). The last period of its activity began in 2006 and continues up to now. In 2006 the brightness increased sharply reacing a maximum. This high state was kept till 2008 and after that time the brightness slowly weakened and reached a minimal value in 2014. Four orbital minima were observed in this period. The depth of the orbital minimum changes. The second minimum is much shallower than the first one and after the second minimum the depth increases, the third minimum is deeper than the second one and the fourth – deeper than the third one (Skopal et al. 2015b). At one time between the first and second minima, since June 2009, satellite components of Balmer lines with a velocity of 300–400 km s^{-1}, indicating high velocity bipolar collimated outflow, appeared in the spectrum and disapeared in the end of 2012. In 2013 their velocity was much lower. A new activity of BF Cyg developed in the end of 2014 and since March 2015 the satellite components restored (Skopal et al. 2015a). The system was in its fifth orbital minimum in March 2016. The satellite components were interpreted in the framework of the model of a collimated stellar wind coming from the outbursting compact companion (Tomov et al. 2014).

In this work, we analyse the optical light curves of BF Cyg during its outburst begun in 2006 with the aim to propose one quantitative interpretation of the change of the depth of the orbital minimum (eclipse) in the framework of the same model. In this way we will trace the evolution of the disc-shaped structure surrounding the compact object which will give useful information for the future development of the models. Our final aim is to suggest a complete interpretation of the behaviour of the optical light of BF Cyg during its outburst after 2006.

For our consideration we will use the ephemeris of Fekel et al. (2001)

$$JD(Min) = 2451395.2^d + 757.2^d \times E,$$

where the eclipse of the compact object, occurring at the time of the inferior conjunction of the giant, is at zero phase.

2. BEHAVIOUR OF THE OPTICAL LIGHT AFTER THE FIRST LIGHT MAXIMUM

After the first maximum in 2006 the light of BF Cyg decreases in different way compared to the next maxima (Fig. 1 in Paper I). Initially the U light falls steeply at about one magnitude. In our view it is most probably due to decrease of the mass-loss rate of the outbursting component which moves the level of the observed photosphere back to the star and redistributes the continuum emission from longer wavelengths towards the UV region. After the initial steeply fall the light begins to decrease more slowly and the basic reason for this decrease is the occultation of the circumbinary nebula. Assuming a decrease of the mass-loss rate, a part of the ejected material can remain within the gravitational potential of the compact object and begin to accrete (Tomov et al. 2014). Observational indication of the increase of the mass-loss rate can be an appearance of absorption component of the spectral lines. During the rise of the light to the second maximum the mass-loss rate was probably increased as proposed by the spectral data of Siviero et al. (2012). Strong P Cyg components of some Balmer and metal lines appeared after 2008 February 10. The light reached its second maximum at the end of 2008. Some time after that the Balmer lines acquired satellite components indicating bipolar collimated outflow (Skopal et al. 2013).

3. INTERPRETATION

A scenario to interpret the spectrum of Z And during its 2000 – 2013 active phase involving two stages of the evolution of the outbursting compact object was presented in the works of Tomov et al. (2012, 2014). It is supposed that there is geometrically thin disc from wind accretion surrounding the compact object in the system in its quiescence and during the first outburst of the active phase. At the end of the first outburst some part of the ejected material accretes again and because of conservation of its initial angular momentum falls into the disc. Thus one extended disc-like envelope covering the disc forms, which locates at a greater distance from the orbital plane than the accretion disc itself. In such a way the second stage of the evolution of the outbursting object begins, which is related to the following outbursts of the active phase. During the following outbursts the envelope can collimate the outflowing material and bipolar collimated outflow (collimated stellar wind) from the outbursting object can form.

The decrease of the depth of the orbital minimum implies an increase of the emission of the uneclipsed geometrical structure of the compact object. We suppose that a thin accretion disc from wind accretion has initially existed in the system BF Cyg and an extended disc-like envelope has formed after that as a result of decrease of the mass-loss rate and accretion of material from the potential well of the compact object in the period between the first and second orbital minima. We suppose that this envelope has collimated the stellar wind later. Some part of it is not eclipsed during the second and following minima determining their smaller depth. The emission of the jets is negligible (Paper I).

Our aim is to calculate the $UBVR_CI_C$ emission of such a model structure and to compare it with the observed residual of the depths of each minimum and the first minimum. If the model emission (the uneclipsed part of the envelope) is close to the observed residual, we can conclude that the formation of a disc-like envelope collimating the stellar wind of the compact object is a possible reason for both the appearance of the satellite components of the spectral lines and change of the depth of the orbital minimum of BF Cyg. After the third minimum the system's brightness fades and the depth of the orbital minimum increases again. We suppose that former is probably due to decrease of the optical flux of the outbursting object and destruction of the envelope and the latter – to destruction of the envelope only.

Gas dynamical modeling shows that a disc from wind accretion with a typical radius of 50 – 60 $R_\odot$ and a mass of 5×10^{-7} $M_\odot$ exists in one binary system with parameters close to those of Z And in its quiescent state (Bisikalo et al. 2002; Tomov et al. 2010, 2011) During the outburst the wind of the compact object "strips" the disc and ejects some part of its mass. At the end of the outburst some part of the ejected mass locates in the potential well of the compact object. After the cessation of the wind it begins to accrete again creating an extended disc-like envelope with a mass smaller than the mass of the initial accretion disc.

The system BF Cyg has parameters very close to those of Z And – orbital period, masses of the components and mass-loss rate of the cool giant (Fekel et al. 2001). Then we suppose that a disc from accretion of a stellar wind with a size and a mass close to that in the system Z And exists in BF Cyg in its quiescent state. During the outburst the newly appeared disc-like envelope should have smaller mass (Paper I).

4. BEHAVIOUR OF THE OPTICAL LIGHT AFTER THE FOURTH ORBITAL MINIMUM

An other activity of the compact object developed at the end of 2014, after the fourth orbital minimum, and the brightness began to grow reaching a maximum in February – March 2015 (Skopal et al. 2015a). The behaviour of the line spectrum reminded that in 2008–2009. In January 2015 the He I and Na I lines had P Cyg absorptions (Munari et al. 2015), and since March satellite Hα components appeared in the spectrum (Skopal et al. 2015a). In this way the Hα profile in July – August 2015 (Fig. 1) was very close to the profile in June – October 2009, containing a red emission satellite component with a velocity of about 300 km s^{-1}, which shows that a collimation mechanism with the same efficiency has probably begun to act in the system in 2015 (Tomov et al. 2018, hereafter Paper III). That is why, by analogy with the period between the first and second orbital minima, we supposed that as a result of a decrease of the mass-loss rate of the outbursting compact object prior to the fourth orbital minimum and accretion of material from its potential well, the disc-like envelope has restored in the time between the fourth and fifth minima on the basis of its remnant from 2014 (see Section 6).

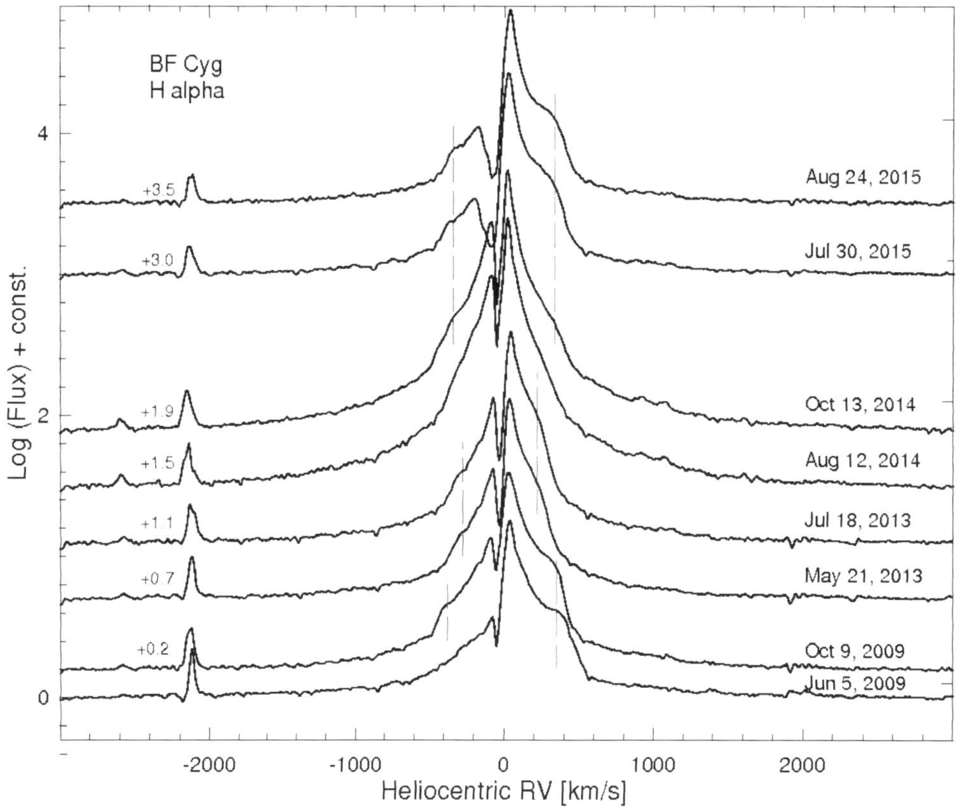

Figure 1: The Hα line profile in 2009, 2013, 2014 and 2015.

5. ANALYSIS OF THE CONTINUUM ENERGY DISTRIBUTION

5.1. UBVR$_C$I$_C$ data used

In order to determine the parameters of the disc-like envelope at the times of the second (Dec. 13, 2009), third (Jan. 23, 2012), fourth (Feb. 11, 2014) and fifth (March 22, 2016) orbital minima it is necessary to have an estimate of the emission measure of the circumbinary nebula of BF Cyg, since it is an upper limit of the emission measure of the disc-like envelope (Tomov et al. 2017, hereafter Paper II). The greatest part of the circumbinary nebula is occulted at the time of the orbital minimum and that is why we estimated its emission measure at the time of its preceding orbital maximum although the nebular emission decreases with time. The brightness maximum on Feb. 26, 2015 is very close to the orbital maximum, which precedes the fifth orbital minimum on March 22, 2016 and we estimated the emission measure for the time of Feb. 26, 2015. In this way we needed in photometric data at the times of the photometric minima and their preceding maxima as well when the occulted part of the nebula is the smallest. We used photometric UBVR$_C$I$_C$ data from the works of Skopal et al. (2012, 2015b, 2017). The continuum fluxes were calculated with use of the calibration data of Mihailov (1973). Unfortunately we were not provided with spectral data in the UBV range and that is why the U flux was not corrected for the energy distribution of BF Cyg in the region of the Balmer jump and, moreover, the UBV fluxes were not corrected for the emission lines. All fluxes were corrected for the interstellar reddening E(B–V) = 0.35 (Skopal 2005) according to Cardelli et al. (1989) and are listed in Table 1.

Table 1: Dereddened fluxes of BF Cyg in units of 10^{-12} erg cm^{-2} s^{-1} A^{-1} with their inner uncertainties at the times of the light minima and maxima under consideration.

Extremum	Date	F_U	F_B	F_V	F_{R_C}	F_{I_C}
Min 1	Dec. 16, 2007	0.720±0.037	0.533±0.015	0.367±0.011	0.330±0.009	0.291±0.009
Min 2	Dec. 13, 2009	1.375±0.065	1.522±0.044	0.883±0.025	0.613±0.017	0.511±0.015
Max 3	Oct. 30, 2010	2.877±0.133	3.004±0.083	1.758±0.048	1.034±0.029	0.629±0.018
Min 3	Jan. 23, 2012	1.316±0.060	1.373±0.038	0.804±0.022	0.543±0.015	0.379±0.010
Max 4	Jan. 07, 2013	2.287±0.106	2.078±0.056	1.334±0.036	0.860±0.024	0.601±0.017
Min 4	Feb. 11, 2014	0.830±0.038	0.754±0.021	0.556±0.016	0.412±0.011	0.346±0.010
	Feb. 26, 2015	3.461±0.159	3.448±0.096	1.841±0.050	1.083±0.030	0.756±0.020
Min 5	March 22, 2016	0.830±0.038	0.907±0.025	0.556±0.016	0.376±0.010	0.330±0.009

5.2. The analysis

To estimate the emission measure of the circumbinary nebula we should analyse the continuum of the system to determine the basic parameters of its components (Paper II). The continuum flux of the giant was determined in the work of Skopal et al (2015b). It is known in this case, and we should determine

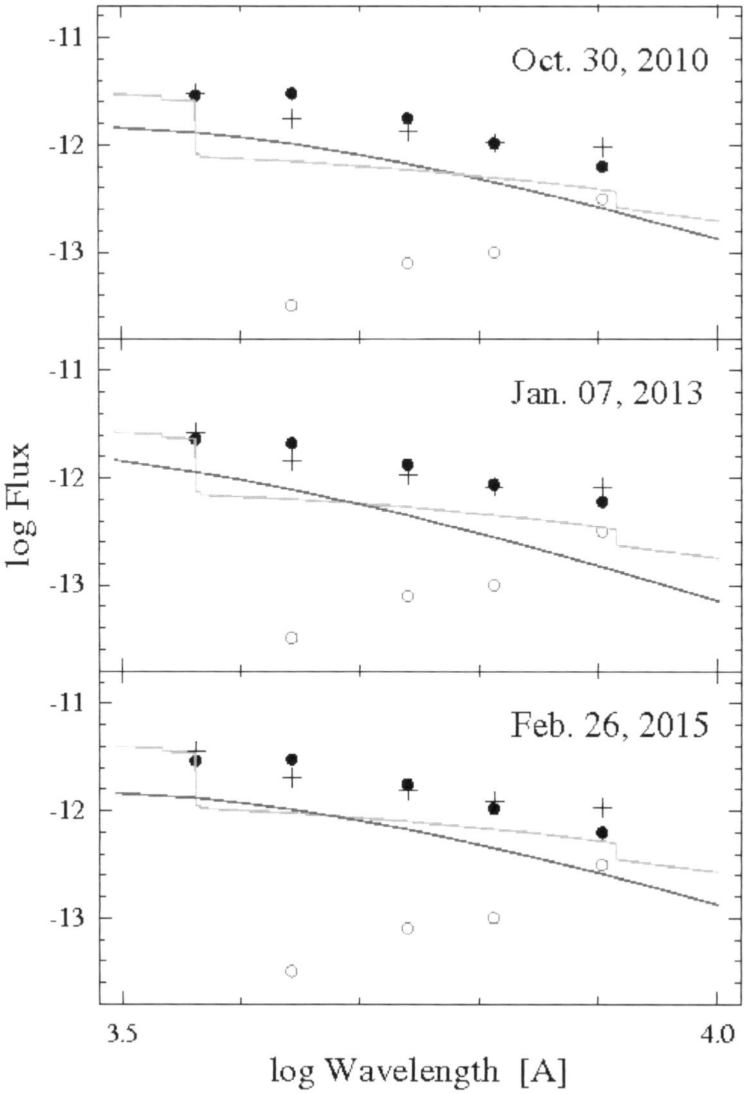

Figure 2: Spectral energy distribution of BF Cyg in the $UBVR_CI_C$ range at the times of third (Oct. 30, 2010) and fourth (Jan. 7, 2013) orbital maxima after 2006 and the light maximum on Feb. 26, 2015. The points indicate the observed fluxes. The lines represent the black body and nebular continua and the circles – the fluxes of the giant. The crosses represent the resulting fluxes.

only the continua of the outbursted compact object and circumbinary nebula. We subtracted the $UBVR_CI_C$ fluxes of the giant from the observed fluxes and approximated their residuals with a nebular continuum and black body emission. In order to approximate the data with a nebular continuum it is necessary to know the dominant ionization state of helium in the nebula. Our unpublished high resolution data do not show presence of the line He II 4686 in the spectrum of BF Cyg. Then, we assume that helium is singly ionized and the nebular continuum is emitted by hydrogen and neutral helium. To calculate the fluxes we used the formula described in Paper I about the continuum flux determined by recombinations and free-free transitions. We used continuum emission coefficients from the paper of Ferland (1980) and the book of Pottasch (1984). We took the arithmetical mean of the hydrogen coefficient on both sides of the Balmer limit at the wavelength of the U band, a helium abundance of 0.1 (Vogel & Nussbaumer 1994) and a distance to the system of 3.8 kpc (Skopal 2005). In this way we approximated the observed fluxes at the times of third (Oct. 30, 2010) and fourth (Jan. 7, 2013) (Table 1) orbital maxima and on Feb. 26, 2015 with a nebular continuum with the same electron temperature of about 30000 K, that was obtained in the work of Skopal et al (2015b) for Oct. 23, 2008 at the time of the second maximum. For the time of the third maximum we obtained an emission measure $n_e^2V = (2.00 \pm 0.15) \times 10^{61}$ cm^{-3}, for the time of the fourth one $n_e^2V = (1.80 \pm 0.12) \times 10^{61}$ cm^{-3} and for Feb. 26, 2015 – $n_e^2V = (2.70 \pm 0.17) \times 10^{61}$ cm^{-3}. In addition the observed fluxes were approximated with black body emission with radius $R = 18.0 \pm 2.0\ R_\odot$ and $T_{eff} = 10000 \pm 500$ K for the time of the third maximum, $R = 10.5 \pm 1.0\ R_\odot$ and $T_{eff} = 13000 \pm 500$ K for the time of the fourth one and $R = 18.0 \pm 2.2\ R_\odot$ and $T_{eff} = 10000 \pm 500$ K for Feb. 26, 2015 respectively. The spectral energy distribution for each time of maximum is shown in Fig. 2. Moreover, we established that the observed fluxes on Oct. 23, 2008 are really well approximated with a warm photosphere with $R = 25\ R_\odot$ and $T_{eff} = 8500$ K as well as with a nebular emission with an emission measure of about 2.6×10^{61} cm^{-3}, obtained in the work of Skopal et al. (2015b), but if the dominant ionization state of helium in the nebula is singly ionized, the continuum of the cool component is better approximated with the giant's photosphere with $R = 150\ R_\odot$ and $T_{eff} = 3400$ K used in the work of Skopal (2005).

6. PARAMETERS OF THE DISC-LIKE ENVELOPE

When determining the parameters of the disc-like envelope we used a binary separation of BF Cyg of 492 $R_\odot$ resulting from the orbital solution of Fekel et al. (2001), a radius of the cool giant $R = 150\ R_\odot$ according to Skopal (2005) and an orbit inclination of 75° according to Skopal et al. (1997) and Fekel et al. (2001). Moreover, we supposed that the shape of the disc-like envelope is close to cylindrical ring (torus).

For the time of the second orbital minimum on Dec. 13, 2009 we accepted an inner radius of the disc-like envelope equal to the radius of the pseudophotosphere of the outbursting compact object on 2008 October 23 according to Skopal et al. (2015b), an outer radius very close to the mean size of the Roche lobe of the compact object, a height of 170 $R_\odot$ and a mass of the envelope of $3.5 \times 10^{-7}\ M_\odot$. Its mean density is thus 7.5×10^{10}cm^{-3} (Table 2). With a radius of the giant of 150 $R_\odot$ about a half of the volume of the disc-like envelope will be visible during the eclipse at orbital phase 0.0 (Paper I). We will calculate the emission of the uneclipsed part of the envelope at this phase and will compare it with the observed residual of the depths of the first and second orbital minima in the photometric bands $UBVR_CI_C$. We accepted an electron temperature in the emitting region T_e = 30000 K equal to the mean temperature in the nebula according to Skopal et al.

(2015b). The fluxes and emission measure of the whole circumbinary nebula, our model envelope and its uneclipsed part at orbital phase 0.0 are presented in Table 3. The emission measure of the nebula was taken from Skopal et al. (2015b) and its fluxes were calculated by us. The quiescent emission measure of BF Cyg at a phase very close to the orbital photometric maximum is 3.1×10^{60} cm^{-3} (Skopal 2005) and on 2008 October 23 – 2.6×10^{61} cm^{-3} (Skopal et al. 2015b). Its increase is thus 2.3×10^{61} cm^{-3}. Then it appears that the emission measure of the envelope of 2.2×10^{61} cm^{-3} is almost equal to this increase.

Table 2: Parameters of the disc-like envelope.

	Minimum 2	Minimum 3	Minimum 4	Minimum 5
R_{in} [$R_\odot$]	25	25	...	18
R_{out} [$R_\odot$]	150	150	...	$93 \div 78$
H [$R_\odot$]	170	130	...	$90 \div 126$
M [$M_\odot$]	3.5×10^{-7}	2.7×10^{-7}	...	0.8×10^{-7}
n_e [cm^{-3}]	7.5×10^{10}	7.5×10^{10}	...	7.5×10^{10}
$n_e^2 V$ [cm^{-3}]	2.21×10^{61}	1.70×10^{61}	$(0.30 \times 10^{61})^a$	0.48×10^{61}

^{a} The emission measure of the uneclipsed part

Table 3: UBVR$_C$I$_C$ continuum fluxes and emission measure of the different regions in the circumbinary nebula. The fluxes are in units of 10^{-12} erg cm^{-2} s^{-1} A^{-1} and emission measure in 10^{61} cm^{-3}.

Min	Emitting region	F_U	F_B	F_V	F_{R_C}	F_{I_C}	$n_e^2 V$
2^a	Whole nebula	2.184	0.913	0.768	0.647	0.491	2.60^b
	Disc-like envelope	1.856	0.776	0.653	0.550	0.417	2.21
	Uneclipsed part	0.928	0.388	0.327	0.275	0.208	1.10
	Residual of the depths	0.655±0.075	0.989±0.046	0.516±0.027	0.283±0.019	0.220±0.017	
	r^c	42	−61	−37	−3	−5	
3	Whole nebula	1.680	0.702	0.591	0.498	0.377	2.00
	Disc-like envelope	1.428	0.596	0.502	0.424	0.320	1.70
	Uneclipsed part	0.714	0.298	0.251	0.212	0.160	0.85
	Residual of the depths	0.596±0.070	0.840±0.041	0.437±0.024	0.213±0.017	0.088±0.013	
	r	20	−64	−43	0	85	
4	Whole nebula	1.512	0.632	0.532	0.448	0.346	1.80
	Disc-like envelope	0.504	0.210	0.178	0.150	0.116	0.60
	Uneclipsed part	0.252	0.105	0.089	0.075	0.058	0.30
	Residual of the depths	0.110±0.053	0.221±0.026	0.189±0.019	0.082±0.014	0.055±0.013	
	r	129	−52	−53	−8	5	
5	Whole nebula	2.268	0.948	0.798	0.672	0.509	2.70
	Disc-like envelope	0.404	0.168	0.142	0.120	0.092	0.48
	Uneclipsed part	0.202	0.084	0.071	0.060	0.046	0.24
	Residual of the depths	0.110±0.053	0.374±0.029	0.189±0.019	0.046±0.013	0.039±0.013	
	r	84	−78	−62	30	18	

^{a}The data for this minimum are from the work of Tomov et al. (2015) and are included in the Table for comparison.
^{b}The data are for 2008 October 23 in the work of Skopal et al. (2015).
^{c}$r = (U - R)/R$ in per cent; U – Uneclipsed part, R – Residual of the depths.

The spectral data of Skopal et al. (2013, 2015b) as well as our spectra in Fig.1 show that the velocity of the satellite components of Balmer lines of BF Cyg was of about 400 km s^{-1} in the period from June 2009 to the end of 2012 and after the beginning of 2013 it decreased. This means that the collimating mechanism preserved until the end of 2012 and changed after that. For this reason we suppose that the parameters of the disc-like envelope at the time of the third orbital minimum have been close to those in the second one, and at least some of these parameters at the time of the fourth minimum have been different from those in second and third ones (Paper II). The results of gas dynamical modeling show that the stellar wind of the compact object destroys its accretion disc – the wind "strips" the disc ejecting some part of its mass and moreover, it can increase its inner radius (Tomov et al. 2010, 2011). The ability of one disc shaped structure to collimate the outflowing gas depends strongly on its inner radius. Since the velocity of the satellite components during the third minimum was the same like in the second one, we adopted the same inner radius and an outer radius very close again to the size of the Roche lobe of the compact object. We adopted a mean density of the envelope of 7.5×10^{10} cm^{-3}, the same as for the second minimum. As for the height of the disc-like envelope, we supposed that it has decreased little, so that with the binary separation, giant's radius and orbit inclination accepted by us, the shadow of the giant passes through the lateral area of the cylinder. When we considered the second orbital minimum we concluded that the eclipsed and uneclipsed parts of the envelope are approximately equal.

We approximated the residual of the depths of the first and third orbital minima in the photometric bands $UBVR_CI_C$ with a nebular emission with T_e = (30000 ± 3000) K and an emission measure of $(0.85 \pm 0.08) \times 10^{61}$ cm^{-3}. This is the emission measure of the uneclipsed part of the envelope at phase 0.0 and according to our conclusion the emission measure of the whole envelope should be $(1.70 \pm 0.16) \times 10^{61}$ cm^{-3} (Table 3). Having the emission measure and mean density of the envelope we obtained its mass $M = 2.7 \times 10^{-7}$ $M_\odot$. Based on this mass and the two radii, we obtained its height of 130 $R_\odot$ (Table 2). With this height and the binary separation, giant's radius and orbit inclination accepted by us, the shadow of the giant will really pass through the lateral area of the cylinder, which means that our supposition is true.

The emission measure of the circumbinary nebula at the time of the third maximum, obtained by us, is $(2.00 \pm 0.15) \times 10^{61}$ cm^{-3}. The quiescent emission measure of BF Cyg at phase close to the orbital maximum is 3.1×10^{60} cm^{-3} (Skopal 2005). It appears that the emission measure of the disc-like envelope is equal to their residual like in the case of the second orbital minimum (Table 3).

An essential feature of the behaviour of the line spectrum of BF Cyg is that from the beginning of 2013 the velocity of the satellite components decreased (Fig. 1), which means that the system has changed, so that its ability to collimate the gas was decreased. Besides the height, it is possible that the inner radius and consequently the mean density of the envelope have changed as well (Paper II). These parameters cannot be estimated from observation and for the time of the fourth minimum we can obtain only the emission measure of the uneclipsed part of the envelope at phase 0.0. We approximated the residual of the depths of the first and fourth orbital minima with nebular emission with T_e = (30000 ± 3000) K and an emission measure of $(0.30 \pm 0.06) \times 10^{61}$ cm^{-3}. This result shows that at the time of the fourth minimum the emission measure of the envelope is not comparable to that of the whole nebula but is much smaller (see Table 3).

The residual of the depths of the first and fifth orbital minima in the photometric bands $UBVR_CI_C$ was approximated with a nebular emission with T_e = (30000 ± 3000) K and an emission measure $n_e^2V = (0.24 \pm 0.07) \times 10^{61}$ cm^{-3}, which should be related to the uneclipsed part of the envelope at phase 0.0. Then the emission measure of the whole envelope amounts to $(0.48 \pm 0.14) \times 10^{61}$ cm^{-3} (Table 3). With the density used by us we obtained a volume of the envelope of 8.53×10^{38} cm^3 and a mass $M \sim 0.8 \times 10^{-7}$ $M_\odot$. The inner radius of the envelope R_{in} is not smaller than the radius of the observed photosphere of the outbursting compact object and we will adopt the radius of this photosphere for the inner radius as we proceeded in the cases of the second and third orbital minima. In regard to the outer radius R_{out} and the height H, we can not determine any of them from observation. The emission measure of the envelope at the times of the second and third orbital minima was great and its mean density to be close to 10^{10} cm^{-3}, the outer radius had to be equal to the radius of the Roche lobe. At the time of the fifth minimum the emission measure is smaller and the outer radius is probably smaller than the Roche lobe. To determine this radius we supposed that the ratio H/R_{in} is within the interval 5–7, since it was in approximately the same interval at the times of the second and third orbital minima, when the Hα profile contained satellite components with the same velocity as in 2015. Using this ratio and an inner radius R_{in} = 18 $R_{\odot}$, we obtained that the outer radius R_{out} and the height H are in the intervals 93–78 $R_{\odot}$ and 90–126 $R_{\odot}$ (Table 2). With these values the shadow of the giant will really pass trough the lateral area of the cylinder and our supposition turned out to be true. The mass and sizes of the disc-like envelope obtained show that it is smaller than at the times of the second and third minima. Its mass and sizes, however, are also close to those predicted by the gas-dynamical modeling (Bisikalo et al. 2002; Tomov et al. 2010, 2011), which means that it can be considered as the accretion structure which collimates the gas at the time of the fifth minimum (Paper III).

7. CONCLUSIONS

We interpreted the variability of the optical brightness of the symbiotic binary BF Cyg during its last outburst begun in 2006 with the evolution of an accretion structure surrounding the outbursting compact object. This structure is related to a disk-like envelope appeared around the initial accretion disk as a result of diminution of the mass-loss rate of the compact object after the first light maximum and accretion of material from its potential well in the time between the first and second orbital minima in December 2007 and December 2009. The uneclipsed part of the disk-like envelope is responsible for the decrease of the depth of the orbital minimum. The envelope is destroyed by the stellar wind of the outbursting object and after the second orbital minimum the depth of the minimum increases again. In the period from the beginning of the outburst to 2014 the optical brightness of the system decreases for two reasons: a diminution of the flux of the compact object because of moving the level of its pseudophotosphere back to the star and redistribution of its continuum from longer wavelengths towards the UV region and destruction of the envelope. In the period between the fourth and fifth orbital minima in February 2014 and March 2016 the disk-like envelope restores from its remnant as a result of accretion of material from the potential well of the compact object again and satellite components with a velocity close to their velocity in 2009–2012 begin to be observed.

We supposed that the shape of the envelope is close to a torus and obtained its parameters for each time of orbital minimum adopting the next parameters of the system: a binary separation of 492 $R_{\odot}$, an orbit inclination of 75° and a radius of the cool giant R = 150 $R_{\odot}$. With these parameters about a half of the volume of the disk-like envelope is visible during the eclipse at orbital phase 0.0. We accepted a mean density of the envelope of 7.5×10^{10} cm^{-3} and determined its inner and outer radii, height and mass. For an inner radius was accepted the radius of the observed photosphere (pseudophotosphere) of the outbursting compact object, which, from its side, was determined from analysis of the continuum energy distribution of the system BF Cyg in the range of the photometric bands $UBVR_CI_C$. The emitted by the uneclipsed part of the envelope $UBVR_CI_C$ fluxes have been calculated and compared with the observed residual of the depths of each orbital minimum and the first one. It turned out that the model fluxes are in satisfactory agreement with the observed ones.

Acknowledgments

This work was partially supported by the Bulgarian National Science Fund of the Ministry of Education and Science under grants DN 08-1/2016 and DN 18-13/2017.

References

Bisikalo D.V., Boyarchuk A.A., Kilpio E.Yu., Kuznetsov O.A.: 2002, *ARep*, **46**, 1022.
Cardelli J.A., Clayton G.C., Mathis G.C.: 1989, *AJ*, **345**, 245.
Fekel F.C., Hinkle K.H., Joyce R.R., Skrutskie M.F. : 2001, *AJ*, **121**, 2219.
Ferland G.J.: 1980, *PASP*, **92**, 596.
Mihailov A.A.: 1973, *Kurs astrofiziki i zvezdnoj astronomii*, Moscow, Nauka, (in Russian).
Munari U., Siviero A., Dallaporta S., et al.: 2015, *ATel* 7013.
Pottasch S.R.: 1984, *Planetary nebulae*, Dordrecht, Reidel.
Siviero A., Tamajo E., Lutz J., Wallerstein G., ANS Collaboration: 2012, *BaltA*, **21**, 188.
Skopal A.: 2005, *A&A*, **440**, 995.
Skopal A., Vittone A.A., Errico L., et al.: 1997, *MNRAS*, **292**, 703.
Skopal A., Shugarov S., Vanco M., et al.: 2012, *AN*, **333**, 242.
Skopal A., Tomov N.A., Tomova M.T.: 2013, *A&A*, **551**, L10.
Skopal A., Sekeráš M., Shugarov S., Pribulla T., Vanko M.: 2015a, *ATel* 7258.
Skopal A., Sekeráš M., Tomov N.A., Tomova M.T., Tarasova T.N., Wolf, M.: 2015b, *AcPol CTU Proc.* **2**, 277.
Skopal A., Sekeráš M., Shugarov S., Shagatova N.: 2017, *ATel* 10086.
Tomov N.A., Bisikalo D.V., Tomova M.T., Kilpio E.Yu.: 2010, *ARep*, **54**, 628.
Tomov N.A., Bisikalo D.V., Tomova M.T., Kilpio E.Yu.: 2011, *AIP Conf. Proc.*, **1356**, 35.
Tomov N.A., Tomova M.T., Bisikalo, D.V.: 2012, *BaltA*, **21**, 112.
Tomov N.A., Tomova M.T., Bisikalo D.V.: 2014, *AN*, **335**, 178.
Tomov N.A., Tomova M.T., Bisikalo D.V., 2015, *AN*, **336**, 690 (Paper I).
Tomov N.A., Tomova M.T., Bisikalo D.V., 2017, *Ap&SS*, **362**, 220 (Paper II).
Tomov N.A., Tomova M.T., Bisikalo D.V., 2018, submitted (Paper III).
Vogel M., Nussbaumer H.: 1994, *A&A*, **284**, 145.

**Proceedings of the XI Bulgarian-Serbian Astronomical Conference
(XI BSAC) Belogradchik, Bulgaria, May 14-18, 2018
Editors: Milcho K. Tsvetkov, Milan S. Dimitrijević and Momchil Dechev
Publ. Astron. Soc. "Rudjer Bošković" No 18, 2018, 159-164**

MILIVOJ JUGIN (1925 – 2013) - GREAT POPULARIZER OF ASTRONAUTICS AND SPACE FLIGHTS

PETAR V. VUCA

Astronomical Society "Rudjer Bošković", Belgrade, Serbia
E-mail: vboba@open.telekom.rs

Abstract. This paper presents the life and work of Milivoj Jugin (Kikinda, August 22, 1925 – January 20, 2013).

Milivoj Jugin was born on August 22, 1925, in Great Kikinda (now the City of Kikinda) to father Maksim, a merchant, and mother Sofija, born Jovanović. He was born in the house at number 31, Dositejeva Street.

Figure 1: The house where Milivoj Jugin lived – photographed by Petar V. Vuca.

He attended primary school and grammar school in Kikinda, finishing it in 1942/1943. He studied at Belgrade Great Technical School, Department of Aviation, which he graduated in 1954. As an aviation engineer he worked on the construction of our well-known jet planes "Galeb"[1] and "Jastreb", as well as "Kraguj", the light piston-engined aircraft. Milivoj Jugin had his artistic side (he

[1] Olivera Stojmirovic, "Milivoj Jugin, constructor of the military aircraft "Galeb" and the live witness of the mision 'Apolo' ", *Glas javnosti,* July 26, 2009.

was a painter, photographer and traveller). He was a doyen of cosmonautics and a great popularizer of space exploration. He gave one of such lectures in the technical school in his native Kikinda. He was an energetic short man, and the father of Miomir, an aviation engineer, and a grandfather to Lara. In his home town Milivoj Jugin had an exhibition of his paintings, where visitors could see a picture painted when he was in grammar school and which was kept by his relative Mirče Jugin in Kikinda. When Milivoj saw the painting, he was surprised and delighted.

Figure 2: The picture Milivoj Jugin painted in grammar school, now kept by his relative from Kikinda – photographed by Petar V. Vuca.

Milivoj Jugin had a great interest in realization of space flights and cosmos exploration before the launch of the Earth's first artificial satellite. He was one of the most active popularizers and promoters of astronautics from the beginning of the cosmic era. He witnessed right on the spot the important moments of man's conquest of space. The climax of his exciting reports from the launching places was the broadcast of Apollo 11 launch from Cape Kennedy. At the moment when the rocket was taking of the ground was shaking, and the roaring sounds of the engines were thoroughly disturbing Cape, Milivoj Jugin uttered his well-known words "horrifying and magnificent". They were his farewell words to Apollo 11 on its flight to the Moon.

"Dear viewers, I am not sure if the roaring sound is reaching you… the ground under our feet is shaking… everything is shaking… This is a horrifying and magnificent sight! The first men have set off to the Moon!", were the words of Milivoj Jugin, a legendary TV commentator and the greatest popularizer of space exploration in the neighbouring countries, which conveyed to millions of Yugoslavs the thrill of the moment when Apollo 11 took off.

"Frivolous are the suspicions that the Apollo crew didn't get to the Moon at all and that Americans faked everything", Milivoj Jugin used to say. "That day I was at Cape Kennedy, having the honour of being on the bus that fetched Armstrong, Aldrin and Collins, and we saw them off to the launching pad; I saw them go to the spacecraft at the top of the "Saturn", the carrier rocket. I saw the launch through, first they entered the Earth orbit, and then set out for the Moon."

"What do you think of the stories about space tourism on the Moon? Don't you have a ticket for the first commercial flight to the Moon, too?" "I do, but I fear that I won't live long enough to travel. And even if I did, I couldn't afford it.""

Figure 3: Aleksandar Đulejić and Milivoj Jugin in TV Belgrade studio during the broadcast of Apollo 14 flight – photo by Slava Čortomić, 1971.

Milivoj Jugin is the author of numerous texts on specialist and other subjects. He wrote eleven books: *Earth's Artificial Satellites* (1960), *Conquest of Cosmos* (1963), *Satellites and Spacecrafts* (1965), *Man and Cosmos* (1969), *Cosmic Technology and Its Application* (1971), *Voyage to Cosmos* (1975), *We Are All Cosmonauts* (1977), *Visiting Universe* (1977), *Military Perspective of Cosmos* (1986), *Cosmos Reveals Secrets* (1997), *Man, Plants, Animals in Cosmos* (1998), and *Eternal Circle* (2000). He delivered a large number (around 500) of popular lectures on astronautics. He was a contributor to many newspapers and magazines. Yugoslav public knew him as the TV Belgrade expert commentator on events in cosmos from 1961. In a live broadcast from Cape Kennedy he commentated on the first launch of men to the Moon in the Apollo 11 spacecraft.

Milivoj Jugin: **"SVI SMO MI KOSMONAUTI"** (**"WE ARE ALL COSMONAUTS"**), Zavod za udžbenike i nastavna sredstva, Beograd, 1995 (Popular science for children)

The author of the book poses a lot of interesting questions commonly raised by young people about the infinity of the universe, a dramatic revelation of its secrets and answers provided by science in recent years, as well as possibilities that can be suggested. This is contained in the book written by engineer Milivoj Jugin with knowledge and great love. He explains clearly and concisely laws of physics, without any mathematics. It is very intelligible to small children. The book demonstrates author's great narrative skills. The book contains answers to the question how the Sun was formed; the question about the Earth's place in the universe; how a satellite flies; traps of the universe; the cosmic carrousel; weightlessness; here and there around the cosmos; we are all cosmonauts; how distant the stars are; whether we are alone in the universe; together in the universe;

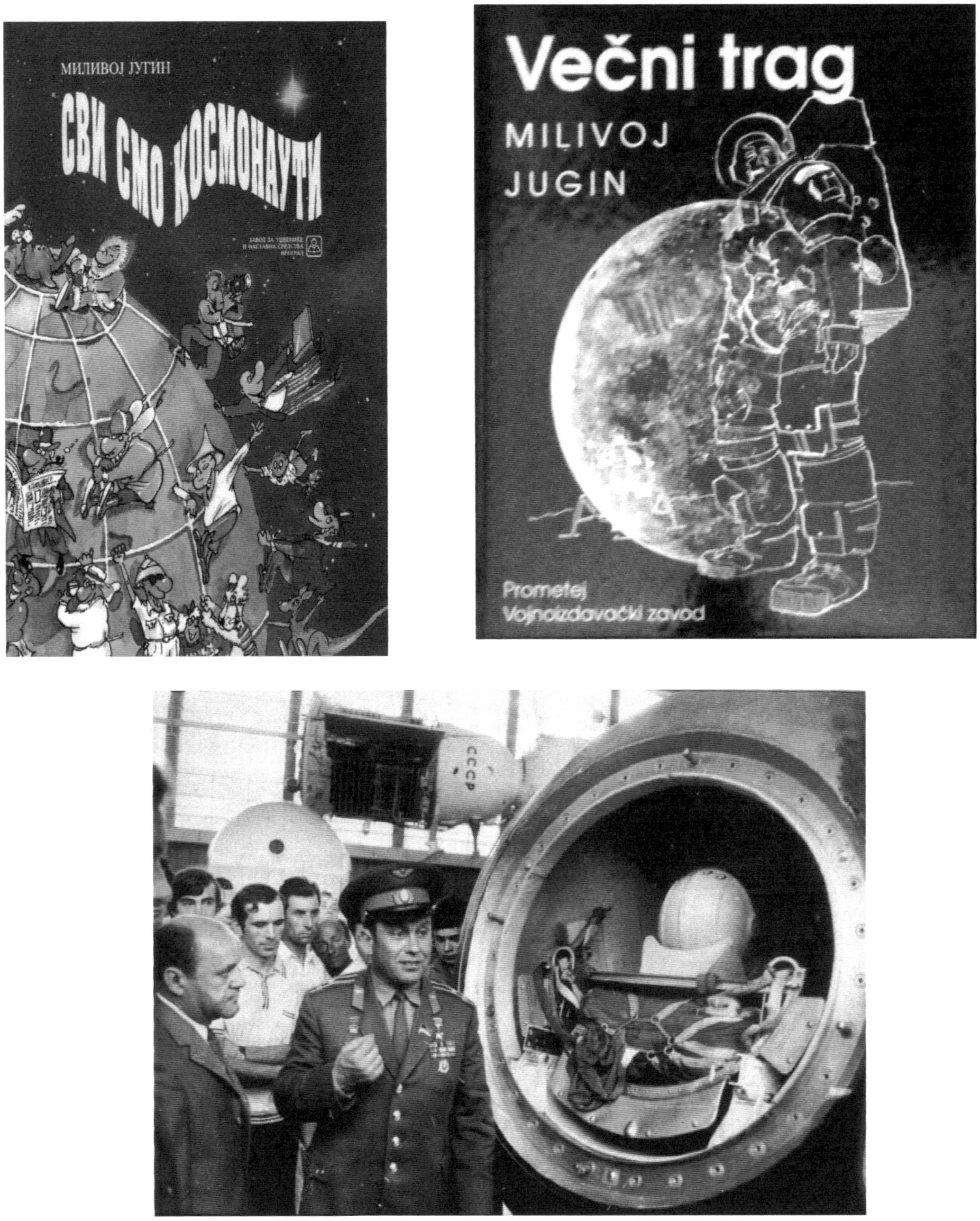

Figure 4: Two books of Milivoj Jugin and the Soviet cosmonaut Pavel Popovich accompanied by Milivoj Jugin during a visit to Belgrade; photographed by Stevan Kragujević (by permission of M. Jugin's daughter Tanja Kragujević)
Milivoj Jugin: **"VEČNI TRAG"** (**"ETERNAL TRACE"**), Prometej, Vojnoizdavački zavod , Novi Sad, 2000.

twelve visitors of the Moon; satellite man; our address in the universe; true stories about Sputniks; about Laika; about astronautic and cosmonautic attractions; spaceships; smart robots that tell men all they have seen and learnt in the form of the most beautiful and exciting fairy tales. Using poetry and jokes, with imagination, Milivoj Jugin takes his readers along the paths passable for children's awareness, he teaches and entertains, and arouses creative curiosity. I recommend this book to all who teach young generations (particularly preschool and junior school teachers). The book is winner of "Neven" literary award.

At the time of writing history, the event of July 16-25, 1969, will hold a special place. That was the moment when Man set out for the Moon, stepped on its surface and returned safely to Earth. During the 10-day Apollo 11 mission people's thoughts throughout the world were focused on achieving its goal. It was the moment of scientific and technological progress.

The central topic of the book "Eternal Trace" is Apollo 11 mission. When the time of the first flight to the Moon passes into oblivion, and when space travel becomes as common as intercontinental air travel, for readers interested in it Milivoj Jugin's book "Eternal Trace" will possess the charm of the few pieces of our literature that have lasting value. Pages of the book make an outstanding contribution of Milivoj Jugin to popularization of science and history of astronautics. The book was written in the beautiful ornate Serbian language. The copies of the book in Serbian are very few. It is richly illustrated with selected photographs and the author's oil paintings. Part of the book contains his personal biography stories, as well as interesting descriptions of his meetings and moments of friendship with famous cosmonauts and astronauts, amongst whom were some of Jugin's personal friends. He brought some of them to his home town Kikinda. The book consists of three parts: **MEMORIES OF THE BEGINNING, MOON DIARY** and **SPACE MEMORIES.**

Figure 5: Astronaut Neil Armstrong, the first man on the Moon (oil painting by Milivoj Jugin).

For years Milivoj Jugin, was the general secretary of the Yugoslav Association of Astronautical and Rocket Organizations (SAROJ) and the president of the Yugoslav Association of Astronautical and Rocket Organizations. He received public recognition and medals for poularization of science. He was a corresponding member of the International Academy of Astronautics (IAA), and attended its annual congresses. He visited USA space centres six times, as well as Cape Canaveral, and attended the launches of Apollo and Space Shuttle spacecraft. Several times he visited Star City of Russian cosmonauts and Baikonur Cosmodrome, where he attended the launch of the Soyuz manned spacecraft.

He was a personal friend and acquaintance of a great number of cosmonauts and astronauts. Milivoj Jugin entertained the famous Russian cosmonaut Viktor Petrovich Savinih. They visited the town of Guča. He took Pavel Popovich, the cosmonaut, for rides in his small Fiat car ("fića"). They were great friends.

Milivoj Jugin was granted permission to attend the Baikonur launch of the new Soyuz TM-5 crew, whose destination was the orbital station "Mir". The launch was successful and nine days later Savinih returned to Earth with his colleagues Solovjev and Aleksandrov.

He professionally organized three international cosmos exhibitions in Belgrade: COSMOS TO PEACE - 1 in 1967; COSMOS TO PEACE – 2 in 1971, which welcomed as guests Popovich, the Soviet cosmonaut, and Thomas Stafford, the American astronaut who commanded the Apollo 10 mission; COSMOS TO PEACE – 3; and also The First International Symposium "Space Exploration and Society" in Belgrade in 1971.

He passed away on 20th January, 2013.[2]

[2] Grujica Ivanović: "Recollections on Milivoj Jugin", *Astronomski magazin*, 19. January 2014.

Poster Papers

Proceedings of the XI Bulgarian-Serbian Astronomical Conference (XI BSAC) Belogradchik, Bulgaria, May 14-18, 2018
Editors: Milcho K. Tsvetkov, Milan S. Dimitrijević and Momchil Dechev
Publ. Astron. Soc. "Rudjer Bošković" No 18, 2018, 167-173

WAVE DENSITY CONFIGURATION VIA THE VORTICITY PATTERNS FORMATION IN ACCRETING BINARY STARS SYSTEMS

DANIELA BONEVA

Space Research and Technology Institute, Bulgarian Academy of Sciences
Acad. Georgi Bonchev St., Block 1, Sofia 1113, Bulgaria
E-mail: danvasan@space.bas.bg

Abstract. We present our theoretical investigations on the local behavior of vorticity formations in accreting flow of binary stars. The vortices are considered as non-single patterns that are moving along with the flow, interacting with the matter there – maintained by gas - dynamical laws. The vortex - flow density relation factor is suggested. Then we study of how much its values are relative to the stability of the vortices themselves and to the entire accretion disc structure.

Further dynamics of the disc matter could result in the situation when a single vortex meets other vortical formations. According to the physical conditions and parameters, these vortices could merge, forming a new bigger and thick composition. In the second case, they could stay as individual vortices, which are held in a cluster and moving as a whole wave configuration.

1. INTRODUCTION

In binary stars with accretion disc, vortices may arise in conditions of tidally interacting flows – which is a common process in this type of stars. After they have appeared, the vortices could propagate globally throughout the disc.

Why we study vortices? As a result of some disc instability, they are responsible for the angular momentum transportation. A configuration of vortices in the regions with accreting flow is one of the most considered mechanisms as efficient angular momentum transportation (Barranco & Marcus 2005) in regions where the magneto-rotational instability (Balbus & Hawle 1998) does not operate. Further they could be considered as sources of brightness variability of the object.

At present, there are many studies in finding the way of vortices appear and behave in the flow. Considering the radial disc structure in a thin disc approximation, it can be followed out by the development of dynamic instability Lovelace et al. (1999). Such instability is the result of the extremum ratio between vortices and surface density Lin (2012). After a series of calculation, vortex formation in weakly self-gravitating discs could be associated with the gap edge.

Li et al. (2000, 2001) have shown that vortices are formed by Rossby waves instability, moving radially and thus transport mass through the disc. This instability, as a source of vortices in the disc, has also been investigated by Meheut (2010). It is usually proceeds in areas with variations in the density values and density accumulation. Vortices can also be generated by a globally unstable radial entropy gradient. The baroclinic instability (Klahr & Bodenheimer 2003, Petersen et al. 2007) has a major role in two-dimensional vorticity formation. We show this in detail in the next section. In our previous papers we obtained a 2D view of the vortex formation (Boneva & Filipov 2012) in the accretion zone flow and their 3D analogue of vortical-like patterns distribution (Boneva 2013).

The research in the current paper is made for close binary stars with tidally interacting flows. We consider the accretion disc flow with already established vortex configuration on the radial direction.

2. VORTEX CONCEPTS: CONDITIONS, MODELS

At a first step, the conditions required for vortex formation and longevity have to be pointed out. Several such models have been introduced in the literature in the last decade. We will show just few of them, which are most related to our study. Bracco et al. (1999) found that if the initial energy of the disturbance is less than about 10^{-3} of the energy in the Keplerian flow, the vorticity fluctuations shear away and the disc returns to its unperturbed velocity profile. In that mean, when we have $E_{init} > 10^{-3} E_{Kepl}$ - for larger initial energies, anticyclonic vortices form and merge, while cyclonic vortices shear out.

If we consider two-dimensional (2D) models of the disc flow structure, one basic source of vorticity, even the only one at some cases - is baroclinicity (Klahr & Bodenheimer 2003). It appears as a term - $\nabla p \times \nabla \rho$ - in the vorticity equation, and is derived by taking the curl of the pressure gradient in the Navier-Stokes equation, where p is the pressure and ρ is the density. The baroclinic term is a source of vorticity in the vorticity equation. The baroclinic term is nonzero when pressure and density gradients are not aligned: $\nabla \rho \times \nabla p \neq 0$. Then, the vortices could be formed.

According to the paper of Petersen et al. (2007), vortex strength increases with the existence and variations of several factors. These are: increasing of the background temperature gradients; warmer background temperatures; larger initial temperature perturbations; and higher resolution.

Godon & Livio (1999) studied vortex longevity with a compressible Navier-Stokes pseudospectral model. It is shown there that the vortex life-time, as measured by the maximum vorticity, is highly dependent on the Reynolds number.

In our current calculations and in the simulations presented here, we imply the conditions, as follows: vortices form when the background temperatures are

$\approx 200K$ and vary radially as $r^{-0.25}$ (r - the relative radius); the initial vorticity perturbations are zero; the initial temperature perturbations are ~5% of the background. When there is a radial variation of the background temperature $T \sim 1/r^n$. We assume that: at $\sim Re = 10^3 \div 10^4$ the maximum vorticity is exponentially growing and decaying in time for about 50 orbital periods (Petersen et al. 2007).

3. VORTEX DYNAMIC: RESULTS

Vortices are not isolated patterns and we consider them as a part of the whole disc structure. The properties of two types of interactions are presented in the next two subsections.

3.1. Vortex – matter interaction

In this first case, we consider vortices as they are moving along with the flow and interacting with the surrounding matter.

Given that we have the vorticity formation with some density ρ_{vort} and radius R_{vort}, which is interacting with the disc flow matter, while moving through it with a variable velocity v_{vort}. The disc matter has a different density ρ_{disc} (which is equivalent to the disc's surface density) and radius R_{disc}. Since the vortex is not an isolated formation and it is moving among the fluid matter, we can write the relative terms for the density, radius and velocity: $\rho_r \approx \dfrac{\rho_{disc}}{\rho_{vort}}$;

$$r_r \approx \frac{R_{disc}}{R_{vort}};\ v_r \approx \frac{v_{flow}}{v_{vort}}.$$

The vortex - flow density relation ρ_r is called "density factor" by Bisikalo et al. (2001). This factor could be responsible for the stability of both: the vortex, as a formation and for the entire accretion disc structure. The results here are intermediate and give some average values.

In our calculations for the average density of the flow, we have: $\rho_{fl} \approx 0.02 \div 0.035 \rho(L_1)$. L_1 is the first (or inner) Lagrangian point. The contrast between the densities of the vortex and the disc could reach the average value of ~ ***1.8*** (see Figure 1).

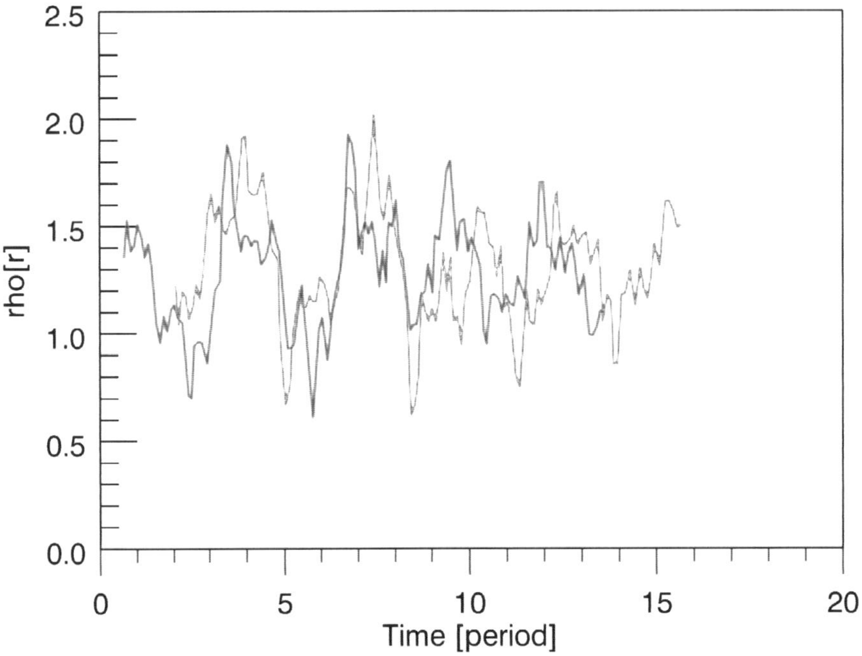

Figure 1: Variations of the density relative term or the "density factor" ρ_r v/s period of rotation. The green and black colors refer to two consecutive stages of calculations.

Following the data, presented at Figure 1, we give the next conditions, related to the values of the "density factor" ρ_r : a) When $\rho_r \approx 1$, we have stable flow and decaying vortices; b) in the case - $1 < \rho_r < 1.8$ - vortices exist in a stable configuration; c) when $\rho_r < 1$ - unstable configuration.

The density factor is not the unique parameter that could define the whole vortices - matter behavior in the disc structure. The physics of matter interaction is complicated and it requires implication of the gas-dynamical laws, deeper analytical and numerical analysis, which were a subject of other paper (Boneva & Filipov 2012).

3.2. Vortex to vortex impacts

In this section, we present one possible way of the vortices interaction activity and environment. The processing of the accretion flow highly influence on the vortex moving conditions.

The matter flux on the radial direction changes in dependence on the velocity v_r and density ρ_r of the moving formation: $\rho_r v_r \sim F_{flux}$ This means that if the density of vortex - like formation grows up, then the flux of matter increases as well. This moving and the growing-up density is in a relation with the accretion dynamic and the accretion rate. Following the denotations of the parameters in section 3.1, the expression for the accretion rate takes the form: $\dot{M} = -2\pi R_{co} v_r \rho_r$, where $\dot{M}$ is the accretion rate; v_r is the radial component of the relative velocity; R_{co} - the distance from the central object. The average value of $\dot{M}$ in close binary stars is: $\dot{M} \approx 10^{-10} \div 10^{-6} M_{\odot} yr^{-1}$.

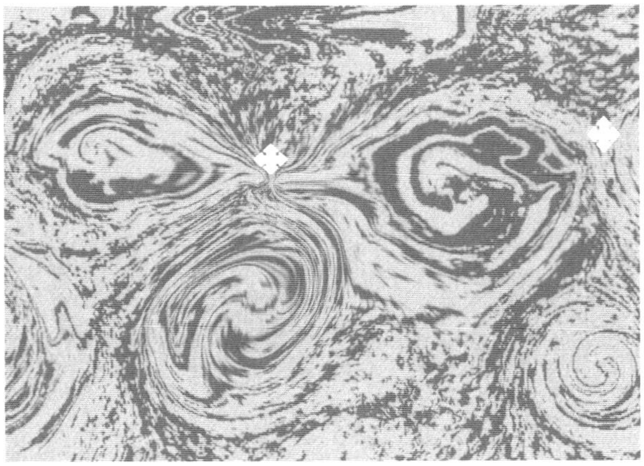

Figure 2: Visualization of the initial step of the vortex – vortex merger process. The contact points are marked with white crosses.

The increasing velocity gives rise to the vortex displacement along to the accretion disc. Than the vortices could contact each other (Figure 2) and their further interaction may lead to the merger of two, three or more vortices in one bigger and ticker vortical formation. Figure 2 gives an illustration of the first contact between the individual ones.

In the case, when the above parameters ($Fflux$, v_r) take lower values, each vortex could stay as a single formation. At some point of time, vortices are attracted to hold each other in a group pattern formation, and influenced by the entire accretion disc flux. Most likely it is affected by the differential rotation and self-gravity in the disc, which are not high enough to induce the merger of the vortices.

Initially, the vortices are formed somewhere locally in the accretion disc flow, they could move as a whole wave pattern throughout the disc on the r, φ direction. Then, the configuration could be transformed to the view, shown at Figure 3.

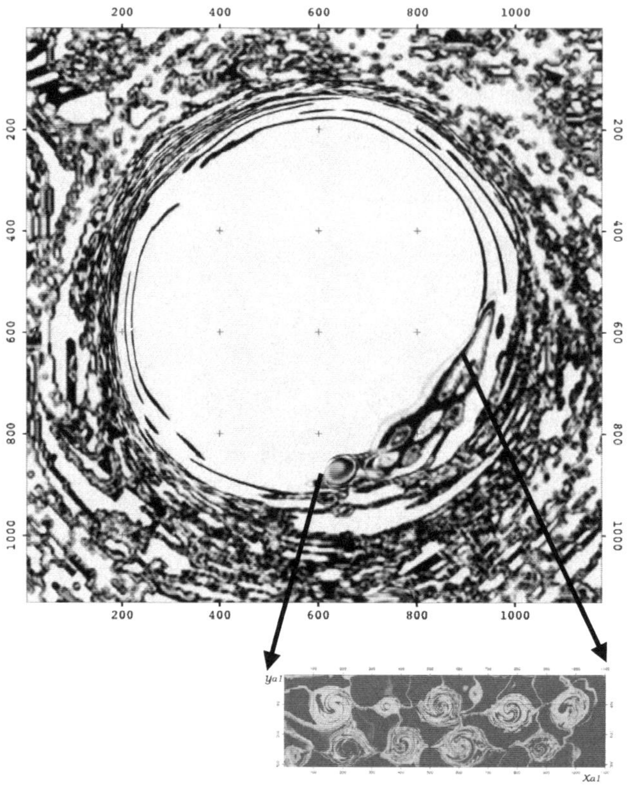

Figure 3: Development of small vortex formations along to the inner part of the accretion disc (Boneva 2017). The simulation presents the possible location of the group pattern, as a snapshot during the image processing. The arrows point the zoomed view of the vortical wave pattern in the calculation frame.

The grids of the Figures 3 correspond to the calculation frame scheme. Their sizes are not related to the boundary conditions, used in the model. The picture is not fully cleared by the side effects and errors, coming from the image processing.

The results in this section are parametric only. Further simulations would involve more data and extend the investigations of this study.

4. CONCLUSIONS

We presented some possible conditions responsible for the development of vortex – like formations, locally in the accretion disc flow, considered for the binary stars systems.

Vortices can be generated by two-dimensional instability as the Rossby wave instability or baroclinic instability, which have been investigated in recent years. In depend on the dominant activity both of the suggested in this paper generation engines and the operating parameters, vorticity could appear as single structures or in a group of several vortices.

In the cases, when the vortices are developed not as a single structure, they could be considered as a chain with tied to each other vortex formations. Further they can evolve as they start merge and form a larger vortex-like or other density structure. Or they could continue their movement throughout the whole disc until their decaying phase ends.

Vortices are initially modeling as local formations, but further under the appropriate physical conditions, they could propagate globally throughout the disc. It is expected for the later calculations that at the higher Reynolds number e-folding time would be at larger orbital periods: >50.

Acknowledgments

This work was initiated in part at Aspen Center for Physics - summer program 2017 and supported by a grant from the Simons Foundation.

References

Barranco, J. A., Marcus, P. S.: 2005, *ApJ* , **623**, 1157.
Balbus, S. A., Hawle, J. F.: 1998, *Rev. Mod. Phys*, **70**, 1.
Bisikalo, D.V., Boyarchuk, A. A., Kil'pio, A. A. Kuznetsov, O. A., Chechetkin, V. M.: 2001, *Astron. Rep.*, **45(8)**, 611.
Boneva, D., Filipov, L.: 2012, astro-ph/1210.2767B.
Boneva, D., Filipov, L.: 2013, *ASP*, **469**, 359-365.
Boneva, D.: 2017, Proceeding of SES'2017, p.79, ISSN 1313 – 3888.
Bracco, A., Chavanis, P. H., Provenzale, A., Spiegel, E. A.: 1999, *Phys. Fluids,* **11**, 2280.
Godon, P., Livio, M.: 1999, *ApJ*, **523**, 350.
Klahr, H., Bodenheimer, P.: 2003, *ApJ*, **582**, 869-892.
Li, H., Finn, J., Lovelace, R., Colgate, S.: 2000, *ApJ*, **533**, 1023.
Li, H., Colgate, S. A., Wendroff, B.,Liska, R.: 2001, *ApJ,* **551**, 874.
Lin, M. K.: 2012, *MNRAS*, **426(4)**, 3211-3224.
Lovelace, R. V. E., Li, H., Colgate, S. A., Nelson, A. F.: 1999, *ApJ*, **513**, 805.
Meheut, H., Casse, F., Varniere, P., Tagger, M.: 2010, *A&A*, **516**, A31.
Petersen, M. R., Julien, K., Steward, G. R.: 2007, *Ap J.*, **658(2),** 1236-1251.

Proceedings of the XI Bulgarian-Serbian Astronomical Conference
(XI BSAC) Belogradchik, Bulgaria, May 14-18, 2018
Editors: Milcho K. Tsvetkov, Milan S. Dimitrijević and Momchil Dechev
Publ. Astron. Soc. "Rudjer Bošković" No 18, 2018, 175-182

SERBIAN-BULGARIAN MINI-NETWORK TELESCOPES AND GAIA-FUN-TO FOR THE PERIOD 2014-2017

GORAN DAMLJANOVIĆ[1], SVETLANA BOEVA[2], OLIVER VINCE[1],
GEORGI LATEV[2], RUMEN BACHEV[2], MILJANA D. JOVANOVIĆ[1],
ZORICA CVETKOVIĆ[1] and RADE PAVLOVIĆ[1]

[1]Astronomical Observatory, Volgina 7, 11060 Belgrade, Serbia
[2]Institute of Astronomy with NAO, BAS, BG-1784, Sofia, Bulgaria
E-mail: gdamljanovic@aob.rs, sboeva@astro.bas.bg, ovince@aob.rs,
glatev@astro.bas.bg, bachevr@astro.bas.bg, miljana@aob.rs, zorica@aob.rs,
rpavlovic@aob.rs

Abstract. About 3000 Gaia Alerts were done by the Gaia Science Alerts group during the last three years (period October 2014 – October 2017). We have observed about 45 objects (about 1650 CCD images) of Gaia-FUN-TO for that period; it is near 15 objects (or 550 CCD images) per year. The standard Johnson BV and Cousins RcIc filters were used, and bias, dark and flat-fielded corrections were applied (plus hot/dead pixels, shutter effect, etc.). The astrometry solution for individual data frames was done in Astrometry.Net (http://astrometry.net), and the Source Extractor was used for the aperture photometry measurements. Finally, the data were sent to the Cambridge Photometry Calibration Server or CPCS (http://www.ast.cam.ac.uk/ioa/wikis/gsawgwiki/index.php/Follow-up) for further calibration. At the end of 2014, the rare object Gaia14aae (GaiaVerif14acp), the deeply eclipsing AM CVn system, was actual and we did our efforts to observe it; as a result, the paper about it (Campbell et al. 2015) was published. In the last two years, another rare object, Gaia16aye – the binary microlensing event, was very interesting and we observed it many times. Also, other rare objects (supernovae, cataclysmic variables, etc.) were observed. To do that, we have used 6 instruments via local cooperation the "Serbian-Bulgarian mini-network telescopes" (established in 2013) and SANU-BAN joint research project "Study of ICRF radio-sources and fast variable astronomical objects" for three-years period (2017-2019, the head is G. Damljanović). In the last two years (2016-2017), the 60 cm Rozhen telescope (Bulgaria) has been under reconstruction, but there is a new one, the 1.4m at ASV (Serbia) from mid-2016 via Belissima project. The first data of the 1.4 m telescope are very good. Some results about the Gaia-FUN-TO are presented, here.

1. INTRODUCTION

The Gaia is a space mission of the European Space Agency – ESA. It is operating since 2014 (astrometrically, photometrically and spectroscopically surveying the full sky). The Gaia-based results are useful for all the relevant scientific communities. It is doing revolution in astrometry, our understanding of the Milky Way galaxy, stellar physics and the Solar system bodies. The main goal of Gaia is to collect the high-precision astrometric data (positions, proper motions, and parallaxes) of sources in the G magnitude range 3 to 21. The Gaia catalogue is an important step in the realization of the Gaia reference frame in future. The second Gaia data release - DR2 (~1.7 billion sources) has been made publicly available on April 2018.

The Gaia scans the sky multiple times, and provides near-real-time photometric data. These data are used to detect some changes in brightness from all over the sky (or appearance of new objects), and the Gaia Science Alerts system produces alerts on some interesting objects. The Gaia Photometric Science Alerts published first alerts at October 2014. Three years after that, the Gaia Science Alerts is among the leading transient surveys in the world. The transients are: supernovae, cataclysmic variables, microlensing events, other rare phenomena. During three years, more than 3000 transients were discovered (until October 2017).

The installation of first telescope (the 60 cm one) at Serbian new site, the Astronomical Station Vidojevica - ASV (of Astronomical Observatory in Belgrade – AOB), was in 2011. At mid-2016 there is a new one, 1.4 m telescope at ASV, via the Belissima project (see http://belissima.aob.rs). Also, we used 4 instruments in Bulgaria (at Belogradchik and Rozhen sites).

2. INSTRUMENTS AND RESULTS

We started the regional cooperation “Serbian-Bulgarian mini-network telescopes” (now, 6 instruments) in 2013. It is in line with the SANU-BAN joint research project “Observations of ICRF radio-sources visible in optical domain” (for the period 2014-2016), and actual one “Study of ICRF radio-sources and fast variable astronomical objects” (2017-2019); head is G.Damljanović. The main information about the instruments is presented in few papers (Damljanović et al. 2014; Taris et al. 2018). It is possible to see the pictures of Schmidt-camera 50/70 cm (Fig. 1), at NAO Rozhen, and 60 cm of ASV with dome (Fig. 2). The NAO BAS means National Astronomical Observatory of Bulgarian Academy of Sciences, and Serbian Academy of Sciences and Arts is SASA (or SANU in Serbian language).

Also, we used the 60 cm and 1.4 m at ASV with other CCDs: the SBIG ST-10 XME (scale=0.23 arcsec and field of view or FoV=8.4x5.7 arcmin using 60 cm at ASV), Apogee Alta E47 (0.45 arcsec and 7.6x7.6 arcmin using 60 cm at ASV), etc.

Figure 1: The Schmidt-camera 50/70 cm (NAO Rozhen, Bulgaria).

Before October 2017, a new aluminization of 2 m Rozhen telescope was finished.

Usually, we did 3 CCD images per filter. The standard bias, dark and flat-fielded corrections are done (also, hot/dead pixels are removed). The Astrometry.Net and Source Extractor are used. The output is supposed to be submitted to the Cambridge Photometric Calibration Server (CPCS) for further calibration. The Johnson-Cousins BVRcIc filters were available.

We collected about 1650 CCD images of the Gaia-Follow-Up Network for Transients Objects (Gaia-FUN-TO, or Gaia Alerts) during about three years (October 2014 – October 2017); or ~550 images per year. It is about 45 objects (near 15 objects per year).

In 2017, we observed 14 objects (about 56% of all objects which we observed) using just the **60 cm telescope at ASV**: Gaia16aye (11 times), Gaia16bnz (2), Gaia17bsu (1), Gaia17bsp (1), Gaia17bsr (1), Gaia17bts (7), Gaia17bxh (1), Gaia17chf (1), Gaia17cgo (1), Gaia17che (1), Gaia17cpa (1), Gaia17cup (2), Gaia17cut (1), and Kojimaevent (1).

With the **1.4 m one at ASV**, we did 5 objects (18%): Gaia16aye (11), Gaia17arv (1), Gaia17asa (1), Gaia17asc (1), and Gaia17aru (1).

With the **60 cm instrument at Belogradchik**, just one object (4%) and it is Gaia17ade (2).

Also, one object (4%) using the **2 m telescope at NAO Rozhen**; it is Gaia16aye (4).

And 5 objects (18%) with **Schmidt-camera 50/70 cm at NAO Rozhen**: Gaia17asc (1), Gaia17arv (1), Gaia17asa (1), Gaia17chf (1), and Gaia17bts (1).

The **60 cm telescope of NAO Rozhen** (0%) was under reconstruction from 2016 until mid-2018, and we expect some data using that instrument.

About some results, the CCD image of **Gaia16aye**, after standard reduction (bias/dark/flat, hot/dead pixels, etc.), is presented (see Fig. 3). The object is marked with cross. That image was made at 19th June 2017 using the 1.4 m ASV telescope with CCD Apogee Alta U42: R-filter, Exp.=40sec, FoV=8.3x8.3 arcmin, binning=1x1, scale=0.243 arcsec . The Gaia16aye was published on Gaia Science Alerts webpage on 9 Aug 2016. The star has a counterpart in the 2MASS as 2MASS 19400112+3007533 towards the Cygnus constellation. That object is the binary microlensing event, and the first discovered in the Northern Galactic Disk. Lot of points (see Fig. 4, light curve of Gaia16aye) are the results of our Serbian-Bulgarian observations. We observed the Gaia16aye 18 times (epochs) during second half of 2016 and 26 times during 2017. At 25th October 2016 we did it using 3 instruments: 50/70 cm Schmidt-camera and 2 m telescope of NAO Rozhen, and 1.4 m instrument of ASV. Together paper about that object is under preparation in the A&A journal (with other co-authors); we hope it will be done, soon.

Figure 2: The 60 cm telescope in its dome (ASV, Serbia).

In line with our cooperation with Dr. Alok Gupta (India) that object was observed 5 nights (21st – 25th November 2016) with the 1.31 m ARIES telescope (Aryabhatta Research Institute of observational sciencES, Manora Peak, Nainital) in the central Himalayan region: λ=79.7E deg, φ=29.4N deg, h=2420 m. The CCD Andor DZ436 was used: 2048x2048 pixels, 13.5x13.5 µm pixel size, the scale is

0.54 arcsec per pixel, FoV=18.5x18.5 arcmin. That instrument is modified R.-C. system Cassegrain; it is Devasthal Fast Optical telescope (DFOT).

Figure 3: The Gaia16aye using the 1.4 m telescope (ASV, Serbia).

Some results of Gaia16aye, in line with observations done at 25th October 2016 (JD=2457687.5), are:

- using **1.4 m telescope at ASV**
  MJD=57686.89422, V=14.949 st.dev.=0.010,
  -\\-.88938, V=14.806 st.dev.=0.010,
  -\\-.89486, r=14.167 st.dev.=0.010,
  -\\-.89241, r=14.108 st.dev.=0.010,
  -\\-.89042, i=13.341 st.dev.=0.010,
  -\\-.89175, V=14.873 st.dev.=0.010,
  -\\-.88990, r=14.116 st.dev.=0.010,
  -\\-.89300, i=13.409 st.dev.=0.010,
  -\\-.89580, i=13.372 st.dev.=0.010,

- using **2 m telescope at NAO Rozhen**
  -\\-.73236, r=14.175 st.dev.=0.005,
  -\\-.72706, B=16.478 st.dev.=0.010,
  -\\-.73198, r=14.177 st.dev.=0.005,
  -\\-.73153, V=14.944 st.dev.=0.005,
  -\\-.73325, i=13.450 st.dev.=0.005,
  -\\-.72902, B=16.451 st.dev.=0.005,
  -\\-.73094, V=14.944 st.dev.=0.005,
  -\\-.73289, i=13.437 st.dev.=0.005,
  -\\-.72531, V=14.965 st.dev.=0.005,
- using **Schmidt-camera 50/70 cm at NAO Rozhen**
  -\\-.81028, r=14.269 st.dev.=0.020,
  -\\-.80504, B=16.538 st.dev.=0.050,
  -\\-.80978, r=14.272 st.dev.=0.020,
  -\\-.80928, V=15.001 st.dev.=0.040,
  -\\-.80347, i=13.457 st.dev.=0.020,
  -\\-.80690, B=16.575 st.dev.=0.060,
  -\\-.80205, r=14.287 st.dev.=0.020,
  -\\-.80599, B=16.567 st.dev.=0.050 .

As we can see from presented values for same filter, the suitable results are close to each other (in accordance with ground-based possibilities) even using different instruments. These magnitudes are transferred from our set of filters (Johnson BV and Cousins RcIc) into another one via the Cambridge Server.

3. CONCLUSIONS

The Gaia satellite (ESA mission) was launched at the end of 2013, and the first astronomical observations were in mid-2014. Since October 2014, the Gaia Photometric Science Alerts (group of Gaia mission) started to publish alerts. Until October 2017 about 3000 alerts (cataclysmic variables, supernovae, candidate microlensing events, etc.) have been issued by the Gaia Science Alerts group.

Using 6 telescopes (via our regional cooperation "Serbian-Bulgarian mini-network telescopes") we observed few objects during the test phase in 2013 and 2014 (Damljanović et al. 2014). Then, we continued the observations of the Gaia-Follow-Up Network for Transients Objects (Gaia-FUN-TO, or Gaia Alerts) from the end of 2014.

As result, we observed about 45 objects until the end of October 2017 (three years of observations); it is near 15 objects per year. About 1650 CCD images were collected (or about 550 ones per year). It was done in Johnson BV and Cousins RcIc filters; usually we did 3 images per filter. The paper (Campbell et al. 2015) about rare object, the eclipsing AM CVn Gaia14aae one, was published. Also, some our results were presented at few conferences.

From mid-2016, we took part in observations of the Gaia16aye (Ayers Rock), binary microlensing event, the first discovered in the Northern Galactic Disk; we hope it will be a published paper about that object, soon. Presented results of thet object, using observations of three instruments (2 m instrument of NAO Rozhen, 1.4 m telescope at ASV, and Schmidt-camera 50/70 cm of NAO Rozhen), are close to each other. That object is just one of Gaia Alerts objects which we observed as very interesting ones. Our plan is to continue with our activities about the Gaia-FUN-TO.

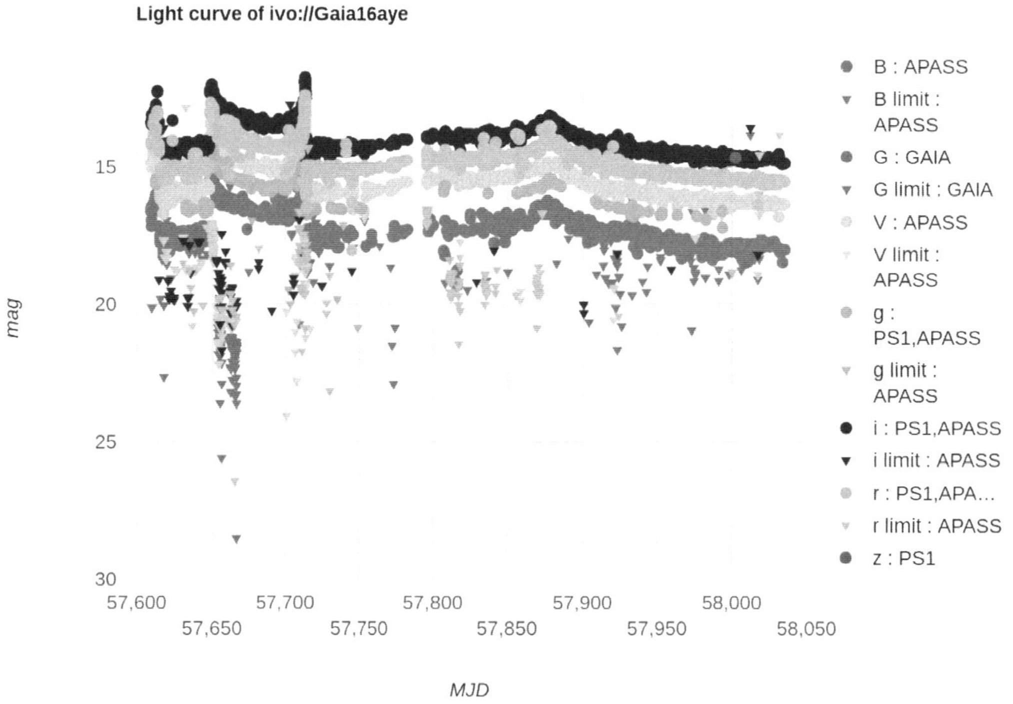

Figure 4: The light curve of Gaia16aye or Ayers Rock (mid-2016 – 9th Oct. 2017).

Acknowledgements

We gratefully acknowledge the observing grant support from the Institute of Astronomy and Rozhen NAO, BAS. This work is part of the projects 176011 “Dynamics and kinematics of celestial bodies and systems”, 176004 “Stellar physics” and 17021 “Visible and invisible matter in nearby galaxies: theory and observations” supported by the Ministry of Education, Science and Technological Development of the Republic of Serbia. Also, we acknowledge ESA Gaia DPAC and Photometric Science Alerts Team

(http://gsaweb.ast.cam.ac.uk/alerts).

References

Damljanović, G., Vince, O., Boeva, S.: 2014, *Serb. Astron. J.*, **188**, 85–93.

Campbell, H.C., et al.: 2018, *MNRAS*, **452**, 1960.

Taris, F., Damljanović, G., Andrei, A., et al.: 2018, *A&A*, **611**, A52.

Proceedings of the XI Bulgarian-Serbian Astronomical Conference (XI BSAC) Belogradchik, Bulgaria, May 14-18, 2018
Editors: Milcho K. Tsvetkov, Milan S. Dimitrijević and Momchil Dechev
Publ. Astron. Soc. "Rudjer Bošković" No 18, 2018, 183-187

DETECTION OF OPTICAL FLICKERING FROM THE SYMBIOTIC MIRA-TYPE BINARY STAR EF AQUILAE

V. V. DIMITROV[1,2], S. BOEVA[1], J. MARTÍ[3],
I. BUJALANCE-FERNÁNDEZ[3], E. SÁNCHEZ-AYASO[3],
G. Y. LATEV[1], Y. M. NIKOLOV[1], B. PETROV[1], K. MUKAI[4,5],
K. A. STOYANOV[1] and R. K. ZAMANOV[1]

[1]*Institute of Astronomy and National Astronomical Observatory, Bulgarian Academy of Sciences, 72 Tsarigradsko Chaussee Blvd., 1784 Sofia, Bulgaria*
[2]*Department of Astronomy, Faculty of Physics, Sofia University "St. Kliment Ohridski", 5 J. Bourchier Blvd., 1164 Sofia, Bulgaria*
[3]*Departamento de Física (EPSJ), Universidad de Jaén, Campus Las Lagunillas, A3-420, E-23071, Jaén, Spain*
[4]*CRESST and X-ray Astrophysics Laboratory, NASA Goddard Space Flight Center, Greenbelt, MD 20771, USA*
[5]*Department of Physics, University of Maryland, Baltimore County, 1000 Hilltop Circle, Baltimore, MD 21250, USA*
E-mail: vdimitrov@astro.bas.bg

Abstract. We performed photometry with a 1 minute time resolution of the symbiotic stars EF Aquilae, AG Pegasi and SU Lyncis in Johnson *B* and *V* band. Our observations of the symbiotic Mira-type star EF Aql demonstrate the presence of stochastic light variations with an amplitude of about 0.25 magnitudes on a time scale of 5 minutes. The observations prove the white dwarf nature of the hot component in the binary system. It is the 11th symbiotic star (among more than 200 symbiotic stars known in our Galaxy) which displays optical flickering. For SU Lyn we do not detect flickering with an amplitude above 0.03 mag in *B* band. For AG Peg, the amplitude of variability in *B* and *V* band is smaller than 0.05 mag and 0.04 mag respectively.

1. INTRODUCTION

Symbiotic stars are long-period interacting binaries, consisting of an evolved giant transferring mass to a hot compact object. Their orbital periods are in the range from 100 days to more than 100 years. A cool giant or supergiant of spectral class G-K-M is the mass donor. The hot secondary component accretes material supplied from the red giant. In most symbiotic stars, the secondary is a degenerate star, typically a white dwarf, subdwarf (Mikołajewska 2003) or neutron star.

There are around 300 known symbiotic stars (Belczyński et al. 2000; Rodríguez-Flores et al. 2014). Among them flickering is detected in only 11 objects (including EF Aql), i.e. in 5% of the cases. On the basis of their infrared properties, the symbiotic stars are divided in three main groups: S-type, D-type, and D'-type (Allen 1982; Mikołajewska 2003). There are about 30 symbiotic stars classified as symbiotic Miras (Whitelock 2003). In three of them flickering is present, i.e. 10% of the objects. It seems that flickering can more often be detected in symbiotic Miras than among S and D'-type symbiotics.

2. OBSERVATIONS

The observations were performed with three telescopes equipped with CCD cameras:

- the 2.0 m RCC telescope of the Rozhen National Astronomical Observatory, Bulgaria (CCD cameras: VersArray 1300 B with 1340×1300 px resolution and 20 μm pixel, and Photometrics CE200A with 1024×1024 px resolution and 24 μm pixel);
- the 50/70 cm Schmidt telescope of Rozhen NAO (CCD camera FLI PL 16803, 4096×4096 px resolution, 9 μm pixel);
- the automated 41 cm Schmidt–Cassegrain telescope of the University of Jaén, Spain (CCD camera SBIG ST10-XME, 2184×1472 px resolution, 6.8 μm pixel; Martí, Luque-Escamilla & García-Hernández 2017).

3. FLICKERING OF EF AQL

EF Aql was identified as a variable star on photographic plates from Königstuhl Observatory (Reinmuth 1925).

Richwine et al. (2005) have examined the optical survey data for EF Aql and classify it as a Mira-type variable with a period of 329.4 d and amplitude of variability > 2.4 mag. Recently, Margon et al. (2016) reported that the optical spectrum shows prominent Balmer emission lines visible through at least H11 and [OIII] λ5007 emission. These emission lines and the bright UV flux detected in GALEX satellite images provide undoubted evidence for the presence of a hot companion as the M primary contributes negligible flux in this bandpass.

Thus EF Aql is classified as a symbiotic star, a member of the symbiotic Mira subgroup (Margon et al. 2016). Rapid aperiodic brightness variations, like the flickering from cataclysmic variables (Bruch 2000), is evident on all of our observations in *B* and *V* bands during 2016 and 2017. The flickering (stochastic photometric variations on timescales of a few minutes with amplitude of a few times 0.1 mag is a variability typical for the accreting white dwarfs in cataclysmic variables and recurrent novae. About the nature of the hot companion in EF Aql, Margon et al. (2016) supposed that the hot source is likely more luminous than a white dwarf, and thus may well be a subdwarf. The persistent presence of minute-timescale stochastic optical variations with the observed amplitude is a strong indicator that the hot component in EF Aql is a white dwarf.

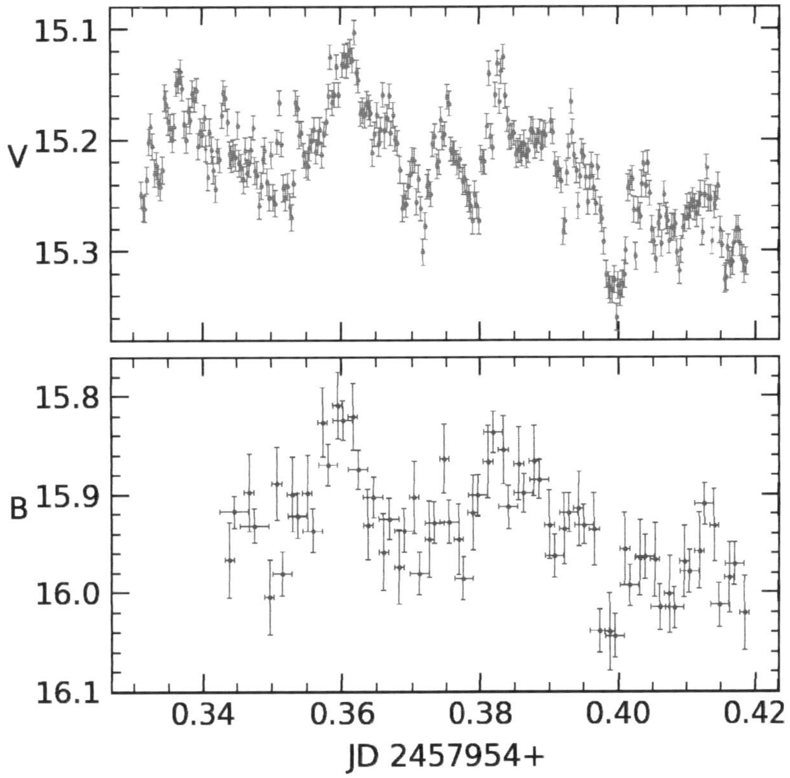

Figure 1: Simultaneous observations of EF Aql in *B* and *V* band. The flickering amplitude in *B* is 0.23 mag, and in *V* – 0.25 mag.

4. OBSERVATIONS OF AG PEG AND SU LYN

AG Peg is the slowest nova ever recorded. Its eruption began in 1855 and continued until 2001. Our observations from September 18, 2017 show that there is no flickering with amplitude above 0.05 in *B* band, and above 0.04 in *V* band.

SU Lyn was observed in X-rays by Mukai et al. (2016), who proposed that this system is a symbiotic star powered purely by accretion onto a white dwarf. The lack of shell burning leads to SU Lyn having very weak symbiotic signatures in the optical. On the night of January 24, 2018 we observed SU Lyn and we did not find any optical variability above 0.03 mag in *B* band.

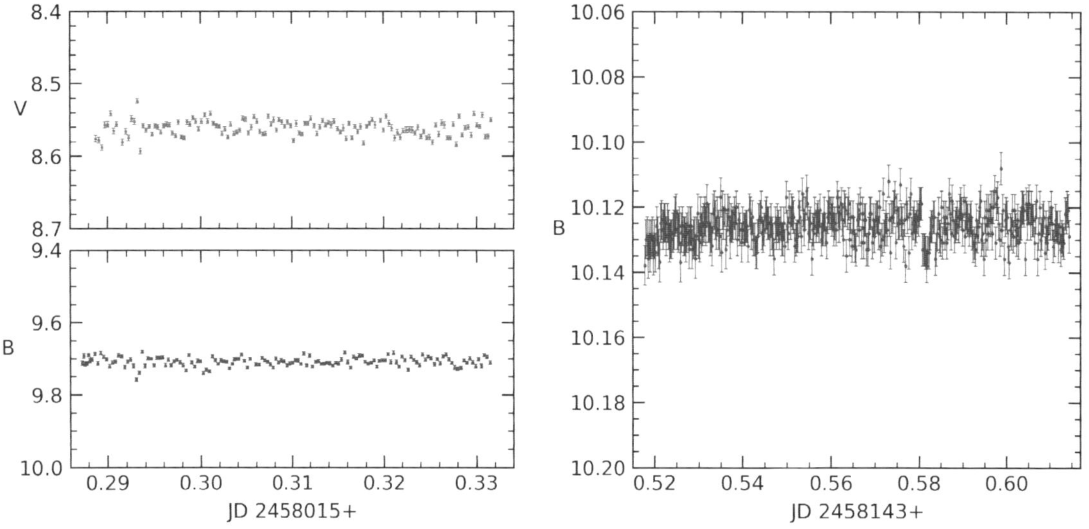

Figure 2: *Left:* simultaneous *B* and *V* band observations of AG Peg. No fast variability is detected. *Right:* observations of SU Lyn in *B* band. No fast variability is detected.

5. DISCUSSION

Most of the flickering-emitting symbiotics are bright X-ray sources. The X-ray emission is divided into five main types: α-type (massive WD with quasi-steady shell burning; Orio et al. 2007); β-type (collision of the winds of the WD and the RG; Muerset, Wolff & Jordan 1997); γ-type (X-rays from a neutron star accretor; Masetti et al. 2007); δ-type (X-rays, emitted from the boundary layer of the accretion disc; Luna et al. 2013); β/δ-type (two X-ray thermal components – soft from the colliding-wind region, and hard from the boundary layer). EF Aql has not been detected in X-rays yet, although it has still not been the subject of a pointed observation with imaging X-ray telescopes.

From the known flickering-emitting symbiotics, there are no α-types, because the energetics of the quasi-steady shell burning are higher than the flickering variability and the flickering becomes undetectable. γ-types are also not detectable, because neutron stars have not been observed as flickering sources.

Among the flickering-emitting symbiotic stars, Z And and o Cet are β-type sources, CH Cyg and MWC 560 are β/δ-type, RT Cru, T CrB and V648 Car are δ-type (Luna et al. 2013). Although RS Oph is considered a δ-type, it shows some peculiar features.

According to Sokoloski, Bildsten & Ho (2001), flickering is much less prevalent in symbiotic stars. The possible explanations are:

- the optical flux is dominated by light from the RG;
- the nebula “washes out” the flickering light;
- lack of an accretion disc;
- shell burning onto the WD surface.

Most probably SU Lyn does not show flickering activity because the flux is dominated by the red giant. In the case of AG Peg, the nova explosion has disrupted the accretion disc and currently the accretion must be spherically-symmetrical.

CONCLUSIONS

We observed optical flickering from the symbiotic star EF Aql with an amplitude of 0.23 mag in B band, and 0.25 mag in V band. The presence of flickering strongly suggests that the hot component is a white dwarf. Our observations of AG Peg and SU Lyn do not show variability above 0.05 mag.

Acknowledgements

This work was partially supported by the Program for career development of young scientists, Bulgarian Academy of Sciences (DFNP 15-5/24.07.2017), by grants DN 08-1/2016, DN 18-10/2017, DN 18-13/2017 (Bulgarian National Science Fund) and AYA2016-76012-C3-3-P from the Spanish Ministerio de Economía y Competitividad (MINECO).

References

Allen, D. A.: 1982, *IAU Colloq. 70: The Nature of Symbiotic Stars*, **95**, 27–42.

Belczyński K., Mikołajewska J., Munari U., Ivison R. J., Friedjung M.: 2000, *Astronomy & Astrophysics Supplement Series*, **146**, 407–435.

Bruch, A.: 2000, *Astronomy & Astrophysics*, **359**, 998–1010.

Luna, G. J. M., Sokoloski, J. L., Mukai, K., Nelson, T.: 2013, *Astronomy & Astrophysics*, **559**, A6.

Martí, J., Luque-Escamilla, P. L., García-Hernández, M. T.: 2017, *Bulgarian Astronomical Journal*, **26**, 91–109.

Masetti, N., Landi, R., Pretorius, M. L., et al.: 2007, *Astronomy & Astrophysics*, **470**, 331–337.

Mikołajewska, J.: 2003, *Symbiotic Stars Probing Stellar Evolution*, **303**, 9–24.

Muerset, U., Wolff, B., Jordan, S.: 1997, *Astronomy & Astrophysics*, **319**, 201–210.

Mukai, K., Luna, G. J. M., Cusumano, G., et al.: 2016, *Monthly Notices of the Royal Astronomical Society*, **461**, L1–5.

Orio, M., Zezas, A., Munari, U., Siviero, A., Tepedelenlioglu, E.: 2007, *The Astrophysical Journal*, **661**, 1105–1111.

Reinmuth, K.: 1925, *Astronomische Nachrichten*, **225**, 385–398.

Richwine, P., Bedient, J., Slater, T., Mattei, J. A.: 2005, *Journal of the American Association of Variable Star Observers*, **34**, 28–35.

Rodríguez-Flores, E. R., Corradi, R. L. M., Mampaso, A., et al.: 2014, *Astronomy & Astrophysics*, **567**, A49.

Sokoloski, J. L., Bildsten, L., Ho, W. C. G.: 2001, *Monthly Notices of the Royal Astronomical Society*, **326**, 553–577.

Whitelock, P. A.: 2003, *Symbiotic Stars Probing Stellar Evolution*, **303**, 41–56.

Proceedings of the XI Bulgarian-Serbian Astronomical Conference (XI BSAC) Belogradchik, Bulgaria, May 14-18, 2018
Editors: Milcho K. Tsvetkov, Milan S. Dimitrijević and Momchil Dechev
Publ. Astron. Soc. "Rudjer Bošković" No 18, 2018, 189-196

A TECHNIQUE FOR AUTOMATIC IMAGE SEQUENCES ALIGNMENT

DIMITAR GARNEVSKI[1], PETYA PAVLOVA[1] and NIKOLA PETROV[2]

[1]Technical University Sofia, branch of Plovdiv, 25 Tsanko Djustabanov Str. 4000 Plovdiv, Bulgaria
[2]Institute of Astronomy and NAO, BAS, 72, Tsarigradsko Chaussee Blvd.1784, Sofia, Bulgaria
E-mail: garnevsky_dm@abv.bg; p_pavlova@gbg.bg; nip.sob@gmail.com

Abstract. Many methods for processing a series of images using algorithms for correlating two images rely on the initial alignment of the objects on the input frames. Tracking the processes in the solar corona requires exact alignment in the positioning of the sun in the analyzed images. The question remains of the greatest importance when the images are obtained using terrestrial instruments - a coronagraph. This is mainly due to the fact that manual positioning of the coronagraphs can lead to a relatively large deviation in the position of the objects in two consecutive frames, which in turn will require manual image alignment during processing or analysis. The main task of this study is to create a technique for arranging two images of the solar corona that are obtained by a coronagraph. The problem relates to process of normalization of the objects view. That means - finding and fixing points inside of the image that allow calculating geometrical features. In addition the progress in registration technology has resulted in improved spatial and time-resolved data. Consequently, increases the amount of information to be processed by increasing the resolution of the images and by reducing the intervals between the records of two consecutive images. That's why we target the implementation of the algorithm by applying advanced image processing technologies, including parallel computing.

1. INTRODUCTION

The mathematics defines the differences between two images of one object to parameters of planar, affine or perspective transformation (Birchfield, 1998). A comparing of these objects supposes a preliminary alignment. The main question at this stage of processing is how to find invariants that allow calculate correctly the transformation parameters. There are two common approaches: feature – based approach and pixel –based approach (Zuliani, 2006),(Nese, 2011). In the first case some geometric features as corners, lines, curves and so on are fixed in the images by their positions and sizes. The values of parameters of transformation are calculated after comparing the chosen features. In the second case the values of parameters are calculated by finding maximal correspondence between whole images, fixed regions or random samples from the model of the object (Zuliani, 2006). The methods for extracting features depend on particular task. For example – in medical image processing (Kumar et all, 2010), (Ching-Wei et al., 2013) are used the both approaches. The free astronomical image processing software –Siril (https://free-astro.org/index.php/Siril), supposes automated alignment for global star alignment (rotation + translation) and translation using DFT (discrete Fourier transformation) centered on an object - more suited for planetary and bright nebula images, and PSF (point spread function) of a single reference star, for deep sky images.

The main task of this study is to create a technique for arranging two images of the solar prominences on the solar limb that are obtained by the Lyot-type coronagraph at NAO Rozhen.

2. THEORY

2.1. Preliminary analysis

The images of the solar corona obtained from the terrestrial instruments consist of two parts - part of the core area of the corona with prominences and a black area of a permanent artificial moon (fig. 1), and the position of the artificial moon variate in the difference images of one sequence.

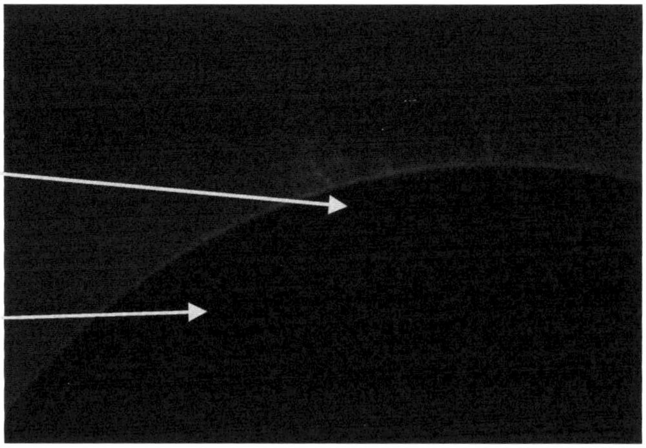

Figure 1: A common view of solar chromosphere and prominences image.

From the point of view of geometry the consecutive frames present planar projections of one variable object. It means that each pair of images could be aligned by parameters of similarity: scaling factor, rotation angle and displacements (Birchfield, 1998). The order of applying the needed transformations is very important and results are real if the center of the object coincides with the center of coordinate system. It means that the first operation must fix a center of the object. The size of the disk of artificial moon for particular instrument is constant. This makes the scaling factor insignificant, and it could be used for calculation the center of coordinate system, displacements of translation and be a center of rotation. Another special peculiarity of the sequences of solar images is dynamics of changes of the analyzed object. It complicates finding the angle of rotation.

2.2. Feature extraction

The simple way to calculate displacements at directions x-dx and y-dy is finding the difference between two positions of one point in consecutive images (fig.2). The technique described below uses a center of circle for this calculation.

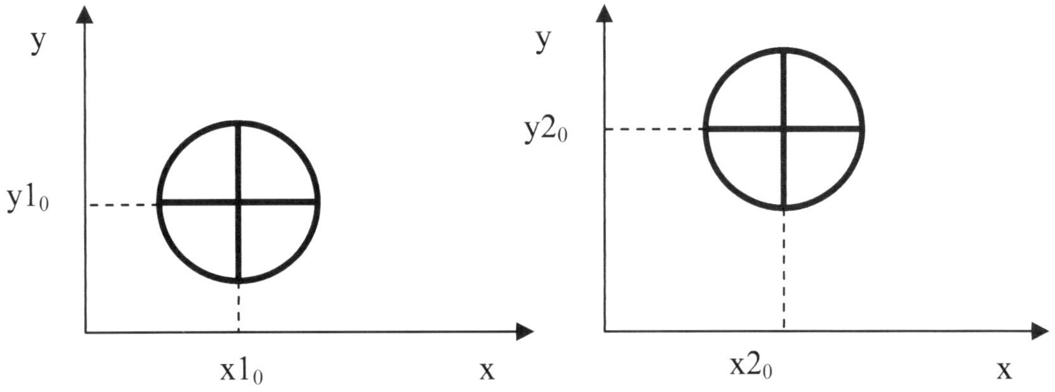

Figure 2: Displacements between two centrums.

$$dx = x2_0 - x1_0 \qquad dy = y2_0 - y1_0 \tag{1}$$

If we have three points $A(x_a,y_a)$, $B(x_b,y_b)$ $C(x_c,y_c)$, the coordinates x_0,y_0 of the centrum of the circle could be calculated by equation (2) (http://mathforum.org/dr.math/):

$$x_0 = \frac{1}{2}.\frac{(y_c - y_b).(x_c^2 + y_c^2 - x_a^2 - y_a^2) - (y_c - y_a).(x_c^2 + y_c^2 - x_b^2 - y_b^2)}{(x_c - x_a).(y_c - y_b) - (x_c - x_b).(y_c - y_a)}$$
$$y_0 = \frac{1}{2}.\frac{(x_c - x_a).(x_c^2 + y_c^2 - x_b^2 - y_b^2) - (x_c - x_b).(x_c^2 + y_c^2 - x_a^2 - y_a^2)}{(x_c - x_a).(y_c - y_b) - (x_c - x_b).(y_c - y_a)} \tag{2}$$

The radius from the points is calculated by equation (3).

$$(x_i - x_0)^2 + (y_i - y_0)^2 = r_i^2 \qquad i=A,B,C \tag{3}$$

The points A, B and C lay on the borders of the artificial moon. That is why the borders in the image have to be outlined. This could be obtained by gradient of the light distribution (Gonzalez et al., 2002). The processing also includes calculation of directional gradients for the central pixel- b_9 under the 3x3 pixels window. The b_i denotes pixels values from the image, and m_i denotes coefficients of gradient operator (4).

$$gr(b_9(x,y)) = \sqrt{gr_x^2 + gr_y^2}$$

$$gr_{x/y} = \sum_{k=1}^{9} m_{i(x/y)}.b_i \tag{4}$$

The gradient image is discrete function and that decrease the accuracy of calculations. That is why the technique applied for finding border points includes tracking the contour by Canny edge detection method (Hoggar, 2012) and calculation the center coordinates by three –pass procedure.

3. IMPLEMENTATION

The algorithm was implemented by parallel processing technique including GPGPU (General Purpose GPU computing) and OpenCV library (http://opencv.org/) for image processing. Figure 3 represents the algorithm used for images alignment.

The algorithm can be described in four common steps:

- Extraction of points over the visible the edge of artificial moon;
- Finding of possible centers of the arc;
- Finding of the real center;
- Alignment of the images.

Alignment of two images is performed with use of affine transformation translation matrix, which can be represented with (5):

$$T = \begin{bmatrix} 1 & 0 & d_x \\ 0 & 1 & d_y \end{bmatrix} \tag{5}$$

Parameters d_x and d_y are calculated as a displacement between the centers of circles in two images.

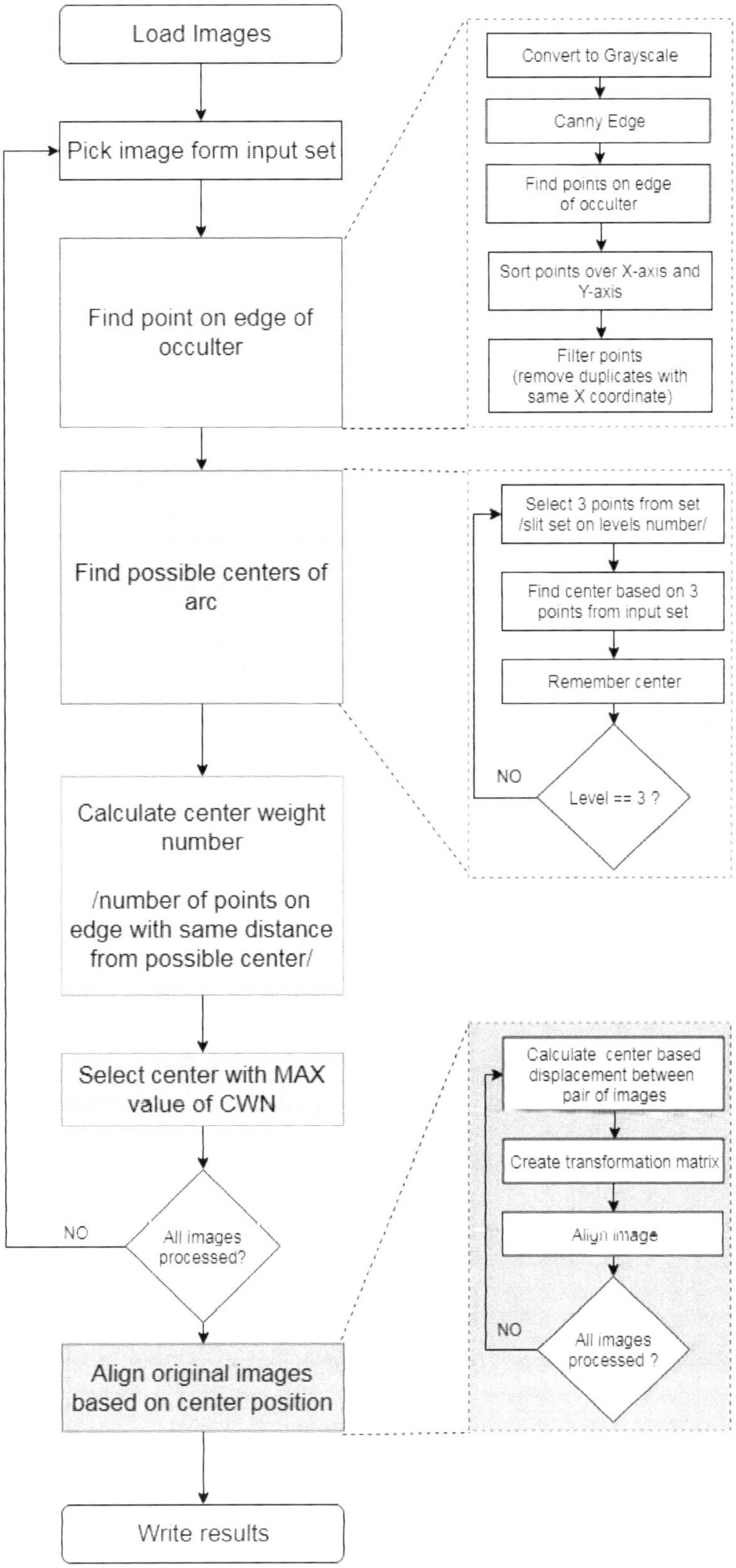

Figure 3: The images alignment algorithm.

4. EXPERIMENT AND RESULTS

For the experiment are used sequences of images of solar corona registered by 15 cm Lyot-coronagraph mounted at NAO Rozhen. The images are 3400 x 2300 pixels. At the images below are represented different stages of the images processing. Color images below describe work of the algorithm responsible for determination of the center of arc based on first, middle and last point from group of points that represents visible part of the occulting disk.

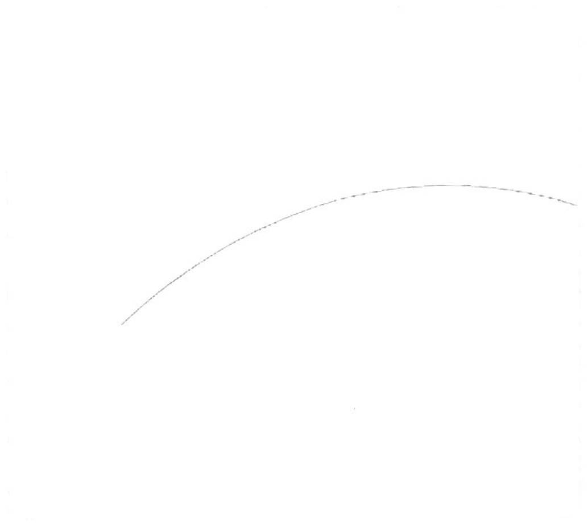

Extracted points of arc

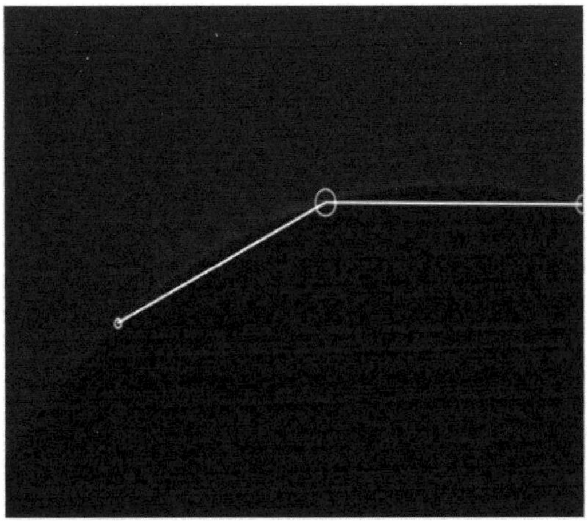

First pass of arc center search (1 sector)

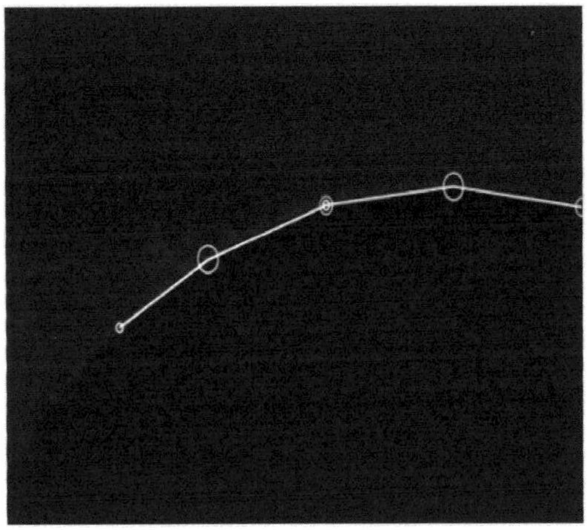

Second pass of arc center search (2 sectors)

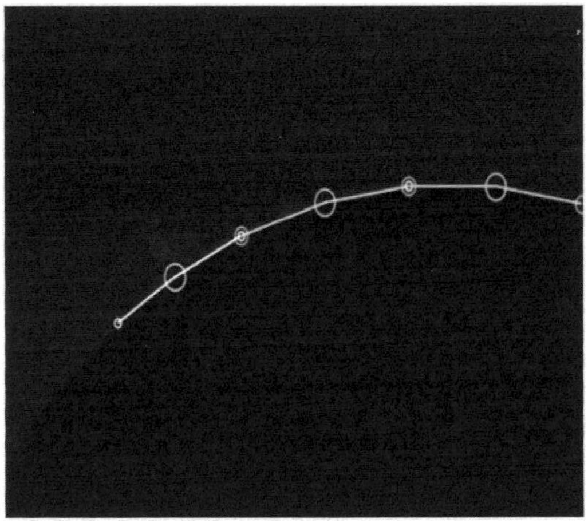

Third pass of arc center search (3 sectors)

Figure 4: Stages of processing.

Figure 5 shows differences of two images before and after alignment.

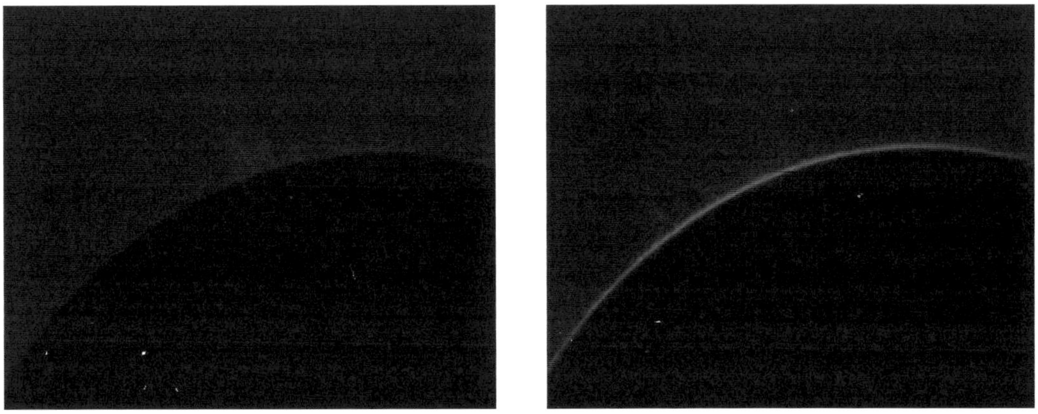

Initial images

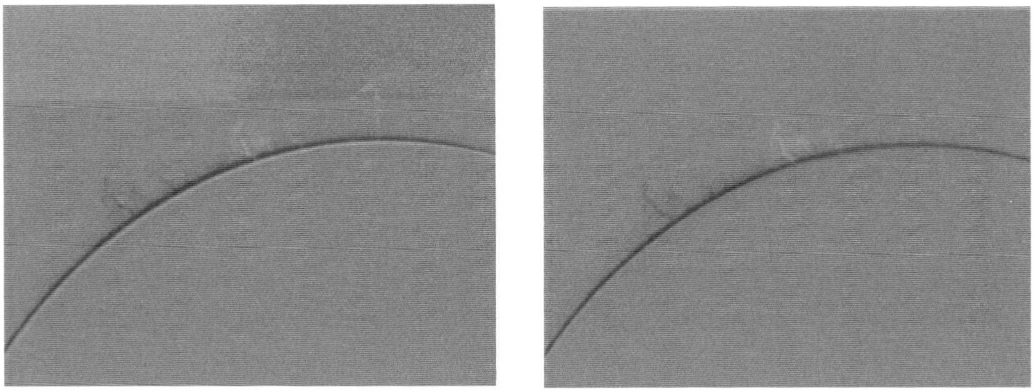

Figure 5: Images of differences before (left) and after (right) alignment.

The calculated center of the first image has coordinates x = 1331 and y = -972 (center of coordinate system is in left bottom corner of the image) and for the second image center is at x = 1339 and y = -961. It means that the second image should be moved to the left with 8 pixels and bottom - with 11 pixels before crop operation for limitation the visible field.

5. CONCLUSIONS

Developed algorithm can effectively handle images that contain less than 50% of the visible part of the occulting disk. In cases with more than 50% of the arc we can use implementation of the Hough transformation to find the circles. Since the lightness and contrast of input images variate in some cases, for optimal results (determination the edge of the occulting disk and elimination the possible false points that are not part of the arc) that insists of additional improvement of these parameters. The full compensation of distortion needs of rotation. However, finding the rotation angle depends on the changes in the picture and makes the result of automatic alignment unreliable in case of unstable solar corona. The problem could be solved using function of differences between two images calculated after step-by-step rotation around the center of the occulting disk.

References

Birchfield S.: 1998, *An introduction to projective geometry (for computer vision)* A.D.

Ching-Wei Wang, Hsiang-Chou Chen: 2013, "Improved image alignment method in application to X-ray images and biological images", *Bioinformatics*, **29,** 1879. doi:10.1093/bioinformatics/btt309.

Gonzalez, R., Woods, R.: 2002, *Digital Image Processing*, 2nd Edition, Prentice Hall.

Hoggar, S. G.: 2012, *Mathematics of digital images. Creation, Compression, Restoration, Recognition*. Cambridge.

Kumar, A., Bandaru, R., Rao, B., Kulkarni, S., Ghatpande, N.: 2010, "Automatic Image Alignment and Stitching of Medical Images with Seam Blending", *IJBBE*, **4**,5.

Nese, M.: 2011 *Automatic Image Alignment, Computational Photography*, Alexei Efros, CMU, Fall.

Open CV (Open Computer Vision) Reference, http://opencv.org/.

Siril: *A free astronomical image processing software*, https://free-astro.org/index.php/Siril

http://mathforum.org/dr.math/

Zuliani, M.: 2006, *Computational Methods for Automatic Image Registration,* PhD thesis, University of California, Santa Barbara.

Proceedings of the XI Bulgarian-Serbian Astronomical Conference (XI BSAC) Belogradchik, Bulgaria, May 14-18, 2018
Editors: Milcho K. Tsvetkov, Milan S. Dimitrijević and Momchil Dechev
Publ. Astron. Soc. "Rudjer Bošković" No 18, 2018, 197-205

FLUX-STABILITY ANALYSIS FOR THE COMPARISON STARS FOR SOME QUASARS IMPORTANT TO ICRF - GAIA CRF LINK

MILJANA D. JOVANOVIĆ, GORAN DAMLJANOVIĆ and OLIVER VINCE

Astronomical Observatory, Volgina 7, 11060 Belgrade, Serbia
E-mail: miljana@aob.rs, gdamljanovic@aob.rs, ovince@aob.rs

Abstract. The Gaia astrometric mission will provide a catalog of about 600 000 quasars and billion stars. Some of these quasars could be the basis of a new optical reference frame. It is necessary to observe a set of common objects (mostly quasars) at both optical and radio wavelengths to link the future Gaia Celestial Reference Frame - Gaia CRF (at optical wavelength) with the ICRF (based on the VLBI observations of sources at radio wavelengths). To establish this link only 10% of the ICRF sources (about 70 ones) are suitable, and because of that this number has been increased by 47 objects (out of the ICRF list) with high astrometric quality. Our observations of 47 candidate sources have been carried out by using three telescopes: two of them with D=0.6m and D=1.4m are located at the Astronomical Station Vidojevica (of the Astronomical Observatory of Belgrade) and the third one with D=2m at the Rozhen NAO BAS (Bulgaria). In this paper we present light curves of a few objects, together with their comparison stars for the period from mid-2016 until now. Also, the results of the flux-stability analysis for the comparison stars are presented here.

1. INTRODUCTION

The European Space Agency mission GAIA (Global Astrometric Interferometer for Astrophysics) was launched in December 2013. The first Gaia data release (Gaia DR1) was made publicly available in September 2016 (Lindegren et al. 2016), and the second one (Gaia DR2) in April 2018 (Gaia Collaboration et al. 2018). The second Gaia release provides complete 5 astrometric parameters (positions, parallaxes, and proper motions) for more than 1.3 billion sources. Plus, the approximate positions for an additional near 0.4 billion mostly faint sources. The reference epoch is J2015.5 (Lindegren et al. 2018).

One of the most important results of the Gaia mission will be a new highly-accurate optical Gaia Celestial Reference Frame (Gaia CRF). That new Gaia CRF will be at the same level of accuracy as the International Celestial Reference Frame (ICRF) that is established in the radio domain by VLBI observations of selected sources (mostly quasars - QSOs). The axes of Gaia CRF will be fixed with respect to distant extragalactic objects. Because of continuity the orientation of the Gaia CRF axes should coincide with the ICRF.

The basic idea to tie the Gaia catalog to the ICRF, is using Gaia observations in optical domain of compact extragalactic ICRF objects with accurate radio positions. Only about 70 sources from ICRF are suitable to establish this link, and because of that this number has increased by 47 sources (out of the ICRF list) with high astrometric quality (Bourda et al. 2011).

There are several papers about some investigation of the flux variability (one of the properties of QSOs) and photocentre motion of QSOs. Taris et al. (2011, 2016), Popovic et al. (2012) have reported numerical values for the photocentre jitter between some tens of μas to some mas. The QSOs with more stable flux should be the base for the link between Gaia CRF and ICRF. Thus, it is necessary to monitor flux stability of QSOs over a longer period of time. In this way, we could determine QSOs with a more stable flux and to eliminate objects with variable photocentre.

2. OBSERVATIONS AND RESULTS

We started with observations of mentioned 47 candidate sources since July 2016 using two telescopes at the Astronomical Station Vidojevica - ASV (of the Astronomical Observatory of Belgrade), and 2 m telescope at the Rozhen NAO BAS (Bulgaria) in accordance with the joint Serbian - Bulgarian research project.

Table 1: The main information about telescopes and used CCD cameras.

Site ASV Telescope Cassegrain D/F = 60/600 cm	CCD camera Apogee Alta E47 pixel array 1024x1024, scale $0.''45$ pixel size $13x13\mu m$ and $FoV = 7.'6x7.'6$
Site Rozhen Telescope Ritchey-Chrétien D/F = 200/1577 cm	CCD camera Andor iKon-L pixel array 2048x2048, scale $0.''17$ pixel size $13.5x13.5\mu m$ and $FoV = 6'x6'$

The characteristics of the 60 cm ASV telescope (with longitude λ = 21.5°E, latitude φ = 43.1°N, altitude h=1140 m) and 2 m Rozhen telescope (λ=24.7°E, φ=41.7°N, h=1730m) with CCD cameras are presented in Table 1. The FoV is field of view. For the other instruments, see in papers (Damljanović et al. 2014, Jovanović et al. 2018). The Johnson-Cousins BVRI filters were available. The standard bias, dark and flat-fielded corrections (plus hot/dead pixels, cosmic rays, etc.) for CCD images are done by using IRAF scripting language.

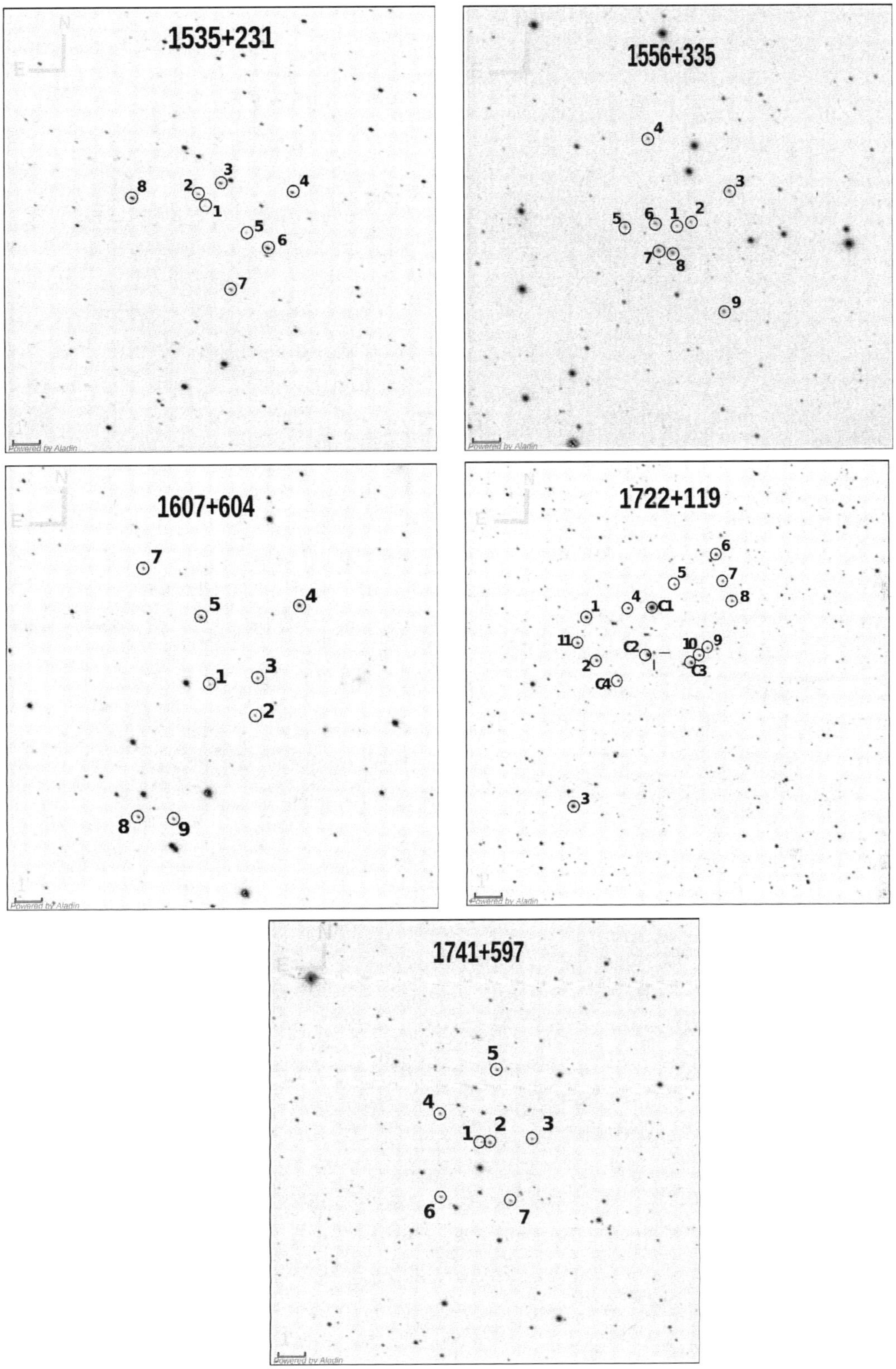

Figure 1: Charts of the fields of objects 1535+231, 1556+335, 1607+604, 1722+119, 1741+597.

Charts of the fields of objects were created using CCD images taken with ASV 60cm and CCD camera Apogee Alta U42 (Fig 1.). In Fig 1. the designation 1 refers to the specific objects from the figure caption and the other numbers refer to comparison stars around the object with exception of 1722+119 where the object is indicated by cross and the other objects are comparison stars.

1722+119. This object is a variable BL Lacertae (BL Lac) with redshift z=0.018. We chose comparison stars with their V and R magnitudes from Doroshenko et al. (2014). They were selected so that some stars are brighter than and some of them are fainter than object. This object has the highest number of comparison stars; we chose 8, out of the 15 stars given in Doroshenko et al. (2014). Comparison stars C2, C3 and 1 are very bright, and were not used for photometry on CCD images where they were saturated.

This object was observed during the period of 776 days, and during that period the V and R magnitudes show some variability (maybe periodicity) in both filters. The average magnitudes in that period are 15.230 (standard deviation, σ=0.130) in V filter and 14.722 (σ=0.180) in R filter. The obtained extremal magnitudes (maximum and minimum values) are: 14.904 and 15.438 in V filter, 14.366 and 14.996 in R filter, respectively. The standard deviation range of comparison stars is from 0.010 (comparison star C2) to 0.045 (star 5) in V filter, and 0.017 (C4) to 0.033 (10) in R filter.

1535+231, 1556+335, 1607+604, and 1741+597. For these objects we chose comparison stars from the SDSS DR14 catalog (https://www.sdss.org/dr14/, Abolfathi et al. 2018). The PSF ugriz-magnitudes were transformed into BVRI using suitable transformations (Chonis and Gaskel 2008).

Comparison stars that we selected for 1535+231 and 1741+597 are much brighter than the objects themselves, but we had to choose stars that better satisfy some criteria (see below). It is not the case with 1556+335 and 1607+604; some of the stars in their vicinity are similar in brightness to these objects, but all stars satisfy criteria (see below) very well. For objects 1535+231 (from 28th April 2017 to 23th April 2018) and 1607+604 (from 4th August 2016 to 15th September 2017), changing CCD cameras on the ASV telescopes caused a change in size of the FoV and, consequently, some of comparison stars could not be able to observe.

Quasar 1535+231, with redshift z=0.46252, during our observational period of 772 days, has average magnitudes of 18.315 (σ=0.166) in V and 18.047 (σ=0.182) in R. The standard deviation of comparison stars is in the range from 0.010 (comparison star 8 in V filter) to 0.039 (star 2 in R filter). The obtained extremal magnitudes of object during that period are: 18.067 and 18.654 in V filter, 17.803 and 18.450 in R filter, respectively. The brightness of object has decreased for about 0.6 magnitude in both filters.

Quasar 1556+335 (redshift z=1.65348) was observed during 776 days. Photometrically, that object is the most stable one, with the average magnitudes of 17.476 (σ=0.040) in V filter and 16.977 (σ=0.036) in R filter. The standard deviation of its comparison stars is in the range from 0.006 (star 7) to 0.033 (star 2) in R filter.

Object 1607+604 is BL Lac with redshift z=0.178. This object was also observed during 776 days. Average magnitudes for this period were: 17.468 with σ=0.087 in V filter, and 17.039 with σ=0.079 in R filter. The lowest standard deviation σ=0.019 has comparison star 4 in R filter, and the highest one σ=0.050 has star 2 in V filter.

Object 1741+597 is also BL Lac with redshift z=0.4. During 775 days, this object had average magnitudes 18.053 (σ=0.206) in V filter and 17.634 (σ=0.252) in R filter; these standard deviations are bigger than others. The magnitude range is from 18.430 to 17.689 in V, and 18.116 to 17.174 in R. It is interesting that the brightness of this object has increased for almost 1 magnitude in both filters. The standard deviation range of comparison stars is from 0.014 (star 4) to 0.036 (star 5) in V filter.

Because of detected changes in magnitudes, objects 1535+231, 1722+119 and 1741+597 are selected to investigate possibility of their intranight changes.

The redshifts were taken from the NASA/IPAC Extragalactic Database - NED (https://ned.ipac.caltech.edu/), and objects type from SIMBAD Astronomical Database.

From SDSS catalog we chose stars following several criteria: not too far from the objects, not too bright or faint stars (with g, r and i magnitudes outside the range 14.5 – 19.5) or not very blue or red (outside the ranges 0.08 < r-i < 0.5 and 0.2 < g-r < 1.4), not variable stars, etc. The PSF ugriz magnitudes of the comparison stars were transformed into the Johnson-Cousins BVRI (Chonis and Gaskel 2008) using equations:

$$V = g - (0.587 \pm 0.022)(g - r) - (0.011 \pm 0.013)$$

$$R = r - (0.272 \pm 0.092)(r - i) - (0.159 \pm 0.022).$$

Obtained V and R magnitudes (Table 2.) were our input values for the Analize tool Photometry in MaxIm DL software. With that tool, which is used for differential photometry, the magnitudes of selected objects and their comparison stars were calculated for each epoch of observation. These output values of the magnitudes of comparison stars for the period of about 2 years were used for flux-stability analysis; the average values of magnitudes are in Table 3. The flux-stability analysis was examined with 3-σ criteria. In accordance with that criteria, we did not detect variability in V and R of comparison stars. In line with standard deviations of stars, the input and output values of suitable magnitudes of stars are close to each other (see Table 2. and 3.). In Fig. 2, the light curves of objects and their comparison stars (output V and R magnitudes) are presented. As we can see, the stars are stable, but some objects show changeable flux during observed period. There are two gaps during the observations due to bad weather conditions, and lack of data for V filter due to technical problems that existed in the period from 7th July to 31th October 2016.

Table 2: Coordinates and magnitudes (V and R) with standard errors of comparison stars for objects 1535+231, 1556+335, 1607+604, 1722+119, 1741+597.

Object	No.	$\alpha_{J2000.0}[^o]$	$\delta_{J2000.0}[^o]$	$V\pm\sigma_V$[mag]	$R\pm\sigma_R$[mag]
1535+231	2	234.314909	23.01831	17.200 ±0.068	16.658 ±0.071
	3	234.300042	23.02486	15.983 ±0.065	15.633 ±0.060
	4	234.251776	23.01917	16.232 ±0.053	15.867 ±0.056
	7	234.293122	22.96096	16.470 ±0.058	15.973 ±0.067
	8	234.359173	23.01592	15.860 ±0.074	15.149 ±0.086
1556+335	2	239.719504	33.39110	17.336 ±0.066	16.850 ±0.073
	3	239.690349	33.40959	16.381 ±0.057	16.095 ±0.056
	5	239.767982	33.38778	16.271 ±0.063	15.916 ±0.060
	6	239.745615	33.39003	16.198 ±0.064	15.825 ±0.061
	7	239.743174	33.37370	15.552 ±0.063	15.188 ±0.060
	8	239.733984	33.37219	15.743 ±0.082	14.897 ±0.104
1607+604	2	242.028820	60.28951	17.068 ±0.059	16.619 ±0.058
	3	242.025260	60.31162	16.864 ±0.055	16.423 ±0.061
	4	241.973520	60.35552	15.195 ±0.054	14.781 ±0.059
	5	242.096377	60.34816	15.630 ±0.064	14.965 ±0.078
	7	242.168538	60.37746	16.856 ±0.053	16.467 ±0.060
1722+119	c2	261.271667	11.86997	13.173 ±0.005	12.570 ±0.006
	c3	261.243750	11.86636	14.078 ±0.012	13.600 ±0.008
	1	261.312083	11.89125	13.445 ±0.009	12.848 ±0.010
	2	261.304583	11.86519	14.823 ±0.008	14.691 ±0.012
	5	261.256667	11.91311	15.873 ±0.010	15.385 ±0.016
	9	261.233333	11.87083	15.809 ±0.008	15.332 ±0.014
	10	261.238750	11.87083	16.142 ±0.011	15.699 ±0.019
	c4	261.289583	11.85344	15.665 ±0.009	15.164 ±0.013
1741+597	2	265.623286	59.75176	15.565 ±0.062	15.204 ±0.063
	3	265.570814	59.75387	16.673 ±0.063	16.314 ±0.062
	4	265.684115	59.76861	16.376 ±0.073	15.795 ±0.073
	5	265.614574	59.79547	16.154 ±0.067	15.704 ±0.064
	6	265.682817	59.71901	16.126 ±0.082	15.684 ±0.085
	7	265.597661	59.71686	16.633 ±0.085	16.124 ±0.091

Table 3: Average magnitudes (V and R) with standard errors of comparison stars for objects 1535+231, 1556+335, 1607+604, 1722+119, 1741+597; period July 2016 - August 2018.

Object	1535+231			1607+604	
No.	V$\pm\sigma_V$[mag]	R$\pm\sigma_R$[mag]	No.	V$\pm\sigma_V$[mag]	R$\pm\sigma_R$[mag]
2	17.217 ±0.025	16.699 ±0.039	2	17.068 ±0.050	16.635 ±0.036
3	16.009 ±0.018	15.661 ±0.034	3	16.886 ±0.033	16.477 ±0.029
4	16.237 ±0.017	15.847 ±0.017	4	15.184 ±0.032	14.760 ±0.019
7	16.447 ±0.023	15.955 ±0.023	5	15.618 ±0.021	14.970 ±0.020
8	15.836 ±0.010	15.138 ±0.022	7	16.848 ±0.027	16.409 ±0.027
	1556+335			1741+597	
2	17.336 ±0.022	16.857 ±0.033	2	15.591 ±0.016	15.251 ±0.029
3	16.364 ±0.023	16.032 ±0.022	3	16.648 ±0.024	16.292 ±0.016
5	16.277 ±0.017	15.892 ±0.022	4	16.383 ±0.014	15.799 ±0.017
6	16.212 ±0.014	15.834 ±0.012	5	16.161 ±0.036	15.711 ±0.018
7	15.556 ±0.014	15.183 ±0.006	6	16.110 ±0.022	15.658 ±0.023
8	15.736 ±0.029	14.926 ±0.008	7	16.599 ±0.022	16.072 ±0.023
	1722+119				
C2	13.174 ±0.010	12.602 ±0.018	5	15.866 ±0.045	15.351 ±0.030
C3	14.080 ±0.015	13.603 ±0.021	9	15.802 ±0.012	15.317 ±0.032
1	13.436 ±0.020	12.822 ±0.028	10	16.132 ±0.025	15.689 ±0.033
2	14.832 ±0.018	14.656 ±0.018	C4	15.654 ±0.021	15.146 ±0.017

5. CONCLUSIONS

In Fig 2. are shown the light curves of objects 1535+231, 1556+335, 1607+604, 1722+119 and 1741+597 with light curves of their comparison stars (output values). For the period of about 2 years, the 3-σ criteria were applied for all listed comparison stars. Some of the objects show changeable flux during observed period, and stars are stable (in line with mentioned criteria). In the measurements of flux of comparison stars any systematic variation was not detected. In line with values of standard deviations, the suitable input and output magnitudes of stars are close to each other. Our next step is to examine magnitude stability of the objects using statistical methods (as F-test, χ^2 one) and to investigate their quasiperiodicities. Also, the presented sets of comparison stars could be used for future investigations of mentioned objects.

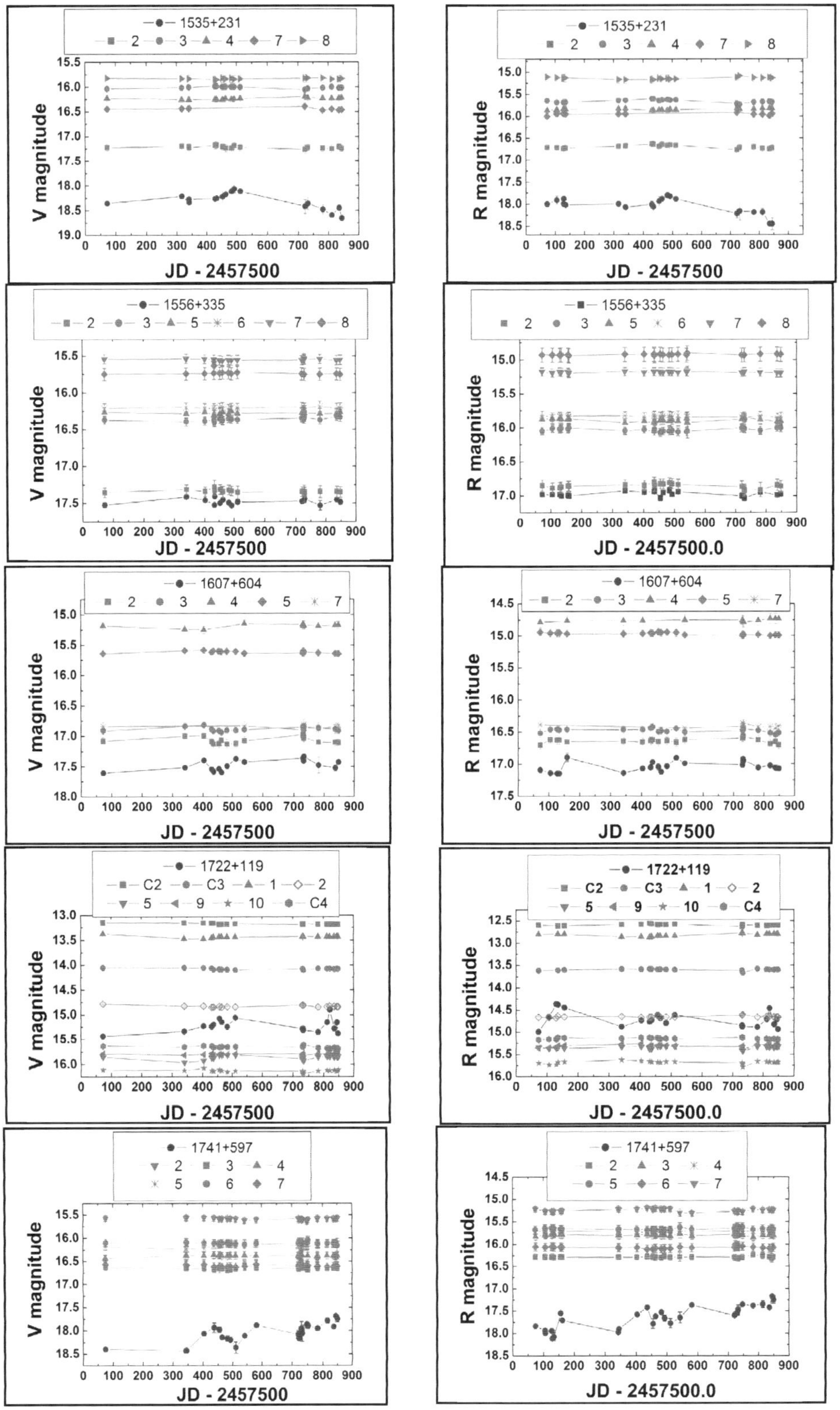

Figure 2: The light curves of V and R magnitudes of the objects and their comparison stars, from July 2016 until August 2018.

Acknowledgment

We gratefully acknowledge the observing grant support from the Institute of Astronomy and NAO Rozhen, BAS, via bilateral joint research project "Study of ICRF radio-sources and fast variable astronomical objects" for the period 2017-2019. This work is part of the projects 176011 "Dynamics and kinematics of celestial bodies and systems", 176004 "Stellar physics" and 17021 "Visible and invisible matter in nearby galaxies: theory and observations" supported by the Ministry of Education, Science and Technological Development of the R. of Serbia.

References

Abolfathi, B., Aguado, D. S., Aguilar, G., et al.: 2018, *ApJS*, **235**, 42.
Bourda, G., Collioud, A., Charlot, P., Porcas, R., Garrington, S.: 2011, *A&A*, **526**, A102.
Chonis, T. S., Gaskell, C. M.: 2008, *The Astronomical Journal*, **135**, 264.
Damljanović, G., Vince, O., Boeva, S.: 2014, Serb. *Astron. J.*, **188**, 85-93.
Doroshenko, V. T., Efimov, Yu. S., Borman, G. A., Pulatova, N. G.: 2014, *Astrophysics*, **57**, 2.
Gaia Collaboration, Brown, A. G. A.,Vallenari, A., Prusti, T., et al.: 2018, *A&A*, **616**, A1.
Jovanović, M. D., Damljanović, G., Vince, O.: 2018, *Publ. Astron. Obs. Belgrade,* **98**, 293 – 296.
Lindegren, L., Lammers, U., Bastian, U., et al.: 2016, *A&A*, **595**, A4.
Lindegren, L., Hernández, J., Bombrun, A., et al.: 2018, *A&A*, **616**, A2.
Popović, L. Č., Jovanović, P., Stalevski, M., et al.: 2012, *A&A*, **538**, A107.
Taris, F., Souchay, J., Andrei, A. H., et al.: 2011, *A&A*, **526**, A25.
Taris, F., Andrei, A., Roland, J., et al.: 2016, *A&A*, **587**, A112.

Proceedings of the XI Bulgarian-Serbian Astronomical Conference
(XI BSAC) Belogradchik, Bulgaria, May 14-18, 2018
Editors: Milcho K. Tsvetkov, Milan S. Dimitrijević and Momchil Dechev
Publ. Astron. Soc. "Rudjer Bošković" No 18, 2018, 207-216

THE SIMEIZ PLATE COLLECTION OF THE ODESSA ASTRONOMICAL OBSERVATORY

SVITLANA KASHUBA[1], MILCHO TSVETKOV[2,3],
NATALYA BAZYEY[1], ELENA ISAEVA[4]
and VALENTINA GOLOVNIA[5]

[1]*Astronomical Observatory of Odessa National University, 65014 Odessa, Ukraine*
[2]*Astronomical Institute with National Astronomical Observatory, Bulgarian Academy of Sciences, 1784 Sofia, Bulgaria*
[3]*Institute of Mathematics and Informatics, Bulgarian Academy Sciences, 1113 Sofia, Bulgaria*
[4]*Institute of Radio astronomy of NAS of Ukraine, URAN-4 observatory, Ukraine*
[5]*Main Astronomical Observatory, National Academy of Sciences of Ukraine. Kyiv, Ukraine*
E-mail: sv-k@onu.edu.ua, milcho.tsvetkov@gmail.com, n.bazyey@onu.edu.ua, isaevaode@gmail.com, golov@mao.kiev.ua

Abstract. Here we describe the Simeiz observatory astronomical plate archive, which is now a part of the Odessa Astrophotonegative Collection. The archive contains about 10000 plates received during period 1908-1953 with Double 120 mm astrograph, which was originally installed in 1908. In the Simeiz archive there are plates taken also at the Kitab International Longitude Station (Uzbekistan) during the World War II where the astronomers from Simeiz were evacuated. The Simeiz plate archive contains observations of small bodies of the Solar system (asteroids), planets and their satellites. Several hundred plates contain images of various objects and phenomena: comets, variable stars, lunar eclipses, etc. Now complete set of plates (including plates obtained in Kitab) are stored in the Astronomical Observatory of Mechnikov Odessa National University at the Mayaki astronomical station in comparatively good conditions. In the period 2014 – 2015 a part of the archive was digitized using the flatbed scanner EPSON Perfection V700. Recently we started work on processing the metadata of the plate envelopes in the machine-readable catalogue compatible with the IVO-WFPDB standards.

1. INTRODUCTION

Our report is devoted to the collection of plates of the Simeiz Observatory. The first plates of this collection were received in 1909. At the end of 1953, the number of plates reached more than 7,000. In 1966, the archive of the Simeiz plates was moved from Crimea to Odessa by agreement between director of Simeiz observatory A.B. Severnyy and director of Odessa observatory V.P. Tsesevich for study variable stars on these plates. Thus, the Simeiz collection became part of the Odessa plate archive, a unique source of information on the long-term behavior of objects (Karetnikov, et al. 1994).

During a difficult period (1909 - 1953), 41 years of observations in Simeiz, a relatively small group of observers showed steadfastness and enthusiasm. The collection of plates was recieved due to astronomers who received world fame: S. I. Belyavsky (1883-1953), G.N. Neuimin (1886-1946), V.A. Albitsky (1891-1952), G.A. Shain (1892-1956), and P.F. Shain (1894 - 1956. Please do not forget names of the founder of the Simeiz Observatory N.S. Maltsov (1849 - 1939) and the organizer and first astronomer of this observatory A.P. Gansky (1870 - 1908).

The main program of observations was the study of small planets of the Solar System. Program also included observations of asteroids, Major planets and their satellites. Several hundred plates contain images of various objects and phenomena: comets, variable stars, lunar eclipses, etc. Having joined the International Organization for the Observation of Small Planets in 1912, the Simeiz Observatory occupied leading place in the world in terms of the number of observations and new objects discoveries before World War II, performing about one-sixth of the whole observational work of organization. So, for example, G.N. Neuimin from 1914 to 1939 published positions of 2791 asteroids, which he observed by himself. Among these, about 250 objects were listed as new. In total, more than ten comets were discovered, 8 of them were given the names of Simeiz scholars; 151 asteroids, including the first opened by Neuimin asteroid, named in honor of Simeiz, place where first amateur observatory in Crimea was founded by N.S. Maltsov. The best known comets that was successfully observed in Simeis are comets Halley (1910), Brooks (1911), Fork (1930), OakRidge (1940), etc. (Karetnikov, et al. 2007).

2. THE SIMEIZ PLATE COLLECTION

A collection of astronomical photographic plates was obtained at the Simeiz Observatory (44° 24' 47"N, 33° 59' 27"E, height - 360 m) (Fig. 1) on the Double 120-mm astrograph with the Unar lens by Carl Zeiss (Fig.2). The analysis of the observation logbooks and its electronic version made it possible to distinguish several observation periods, which are separated in eight separate archives according to the specification of the Database in Sofia (Tsvetkov, 2006). In general, the whole collection contains about 8 thousand plates.

Figure 1: Simeiz Observatory (photo 1908).

It was founded by N. S. Maltsov as an amateur observatory, but was later associated as a branch of the Pulkovo Observatory with the participation of the astronomer A. P. Gansky.

In the evacuation period (1942-1944), during World War II, observations were made at the Bredikhin Double astrograph in Kitab, Uzbekistan.

Characteristics of the Simeiz Double astrograph are given in Table 1.

Table 1: Date of observations is from JD 2418405 to 2434725 (1909-1953).

Type of objective	No. of cameras	D, mm	F, mm	Plate size, mm	Field of view, °	Lim. Photographmag, ^{m}	FoV center shift from guide star, °	Emulsion types	Exposition
UNAR	1 & 2	120	600	130 x 180	12 x 16	~ 15 (pg)	0	> 10	up to 2^h

The centers of the plate field were determined by the position of small planets, comets, and their expected area of presence. Most observations of small bodies was fulfilled according to the method of Metcalf: the method of intermittent guiding of an object, with exposures up to 1.5 - 2 hours. The coordinates are indicated on the envelopes of the plates. In Fig. 3 we gave the observational statistics of 1910 as an illustration of the seasonal distribution of observation nights during the year. Distribution of plates with years is shown in Fig. 4.

Figure 2: Double 120 mm astrograph with Unar lens by Carl Zeiss (established in 1908).

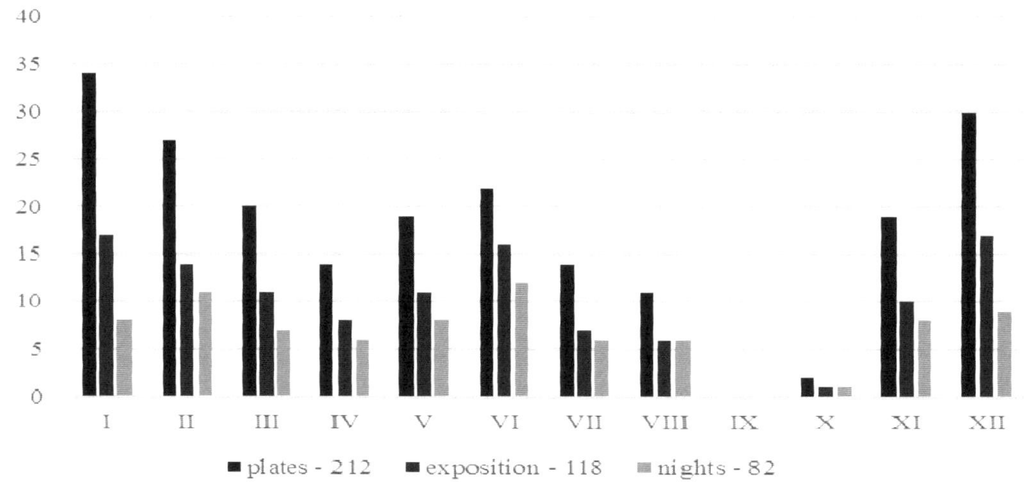

Figure 3: Observation statistics in Simeiz in 1910.

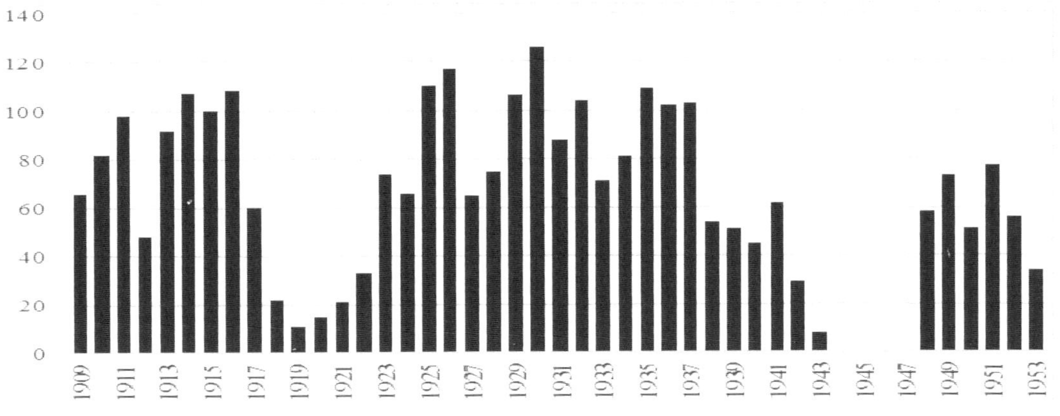

Figure 4: Observation statistics in Simeiz from 1909 to 1953 (Andrievsky, et al. 2007).

During more than 40 observational years more than 10 types of astronomical emulsions were used. Unfortunately, we do not have much information about types, recipes, methods of processing etc.

3. PRESENT STATE OF THE SIMEIZ PLATES COLLECTION

Now all plates of the Simeiz collection are stored in the Astronomical Observatory of Mechnikov Odessa National University at suburban Mayaki astronomical station (40 km from Odessa).

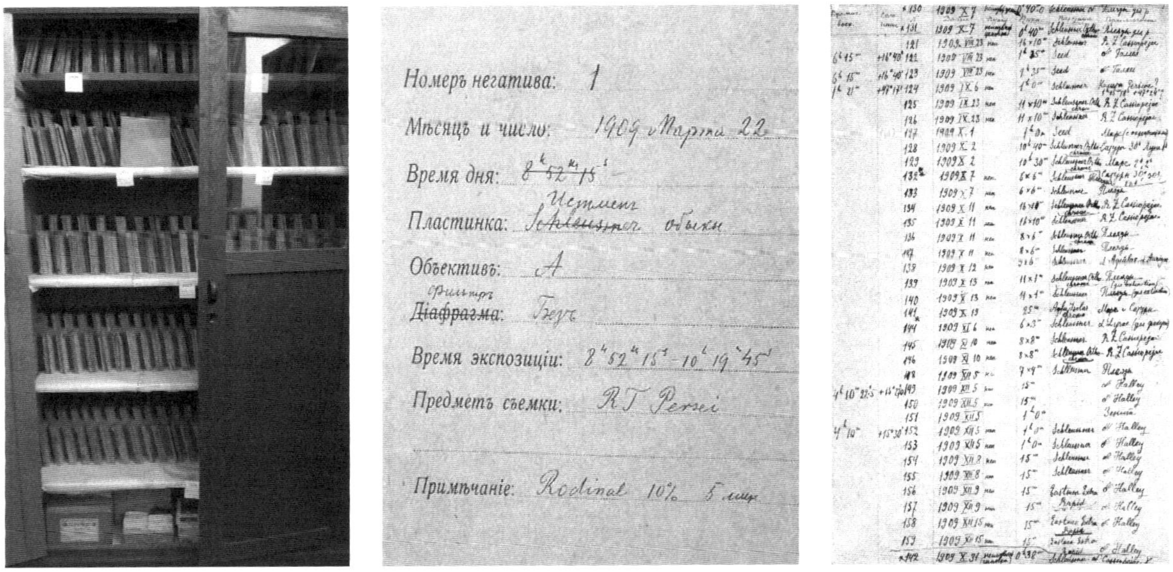

a b c

Figure 5. a: Wooden cupboard with the Simeiz plates, **b:** envelope of the first plate in collection, **c:** Fragment of an observational logbook.

The plates are packed in paper envelopes and stored in time ascending order in 4 wooden cabinets in an isolated from the whole collection room in the administrative building of the observation station. Storage conditions are close to maximum allowable: with low air circulation, seasonal changes in air temperature in the range +3 - +30°C and non-controlled humidity of air. Also re-recorded logs are stored in the glass library as the originals are absent.

4. DIGITIZATION OF THE SIMEIS COLLECTION OF PLATES

Since 2013, the project of digitizing the Odessa photographic plates archive has been launched. Plates in small batches were transported from Odessa to Kiev (the Main Astronomical Observatory of National Academy of Science of Ukraine). About 400 plates of the Simeiz collection were digitized on the scanner Epson Expression 10000XL for one year. Images were digitized as 16-bit grayscale FITS with a resolution 1200 dpi. Size of one image is about 400MB.

Images integrated in the combined digital archive of Ukrainian Virtual Observatory Joint observation Data Archive (JDA). To date, the JDA contains 600 metadata records and more than 400 digitized images of selected plates of the Simeiz collection (1951 - 1953) with low resolution (for preview) (Vavilova, et al. 2012).

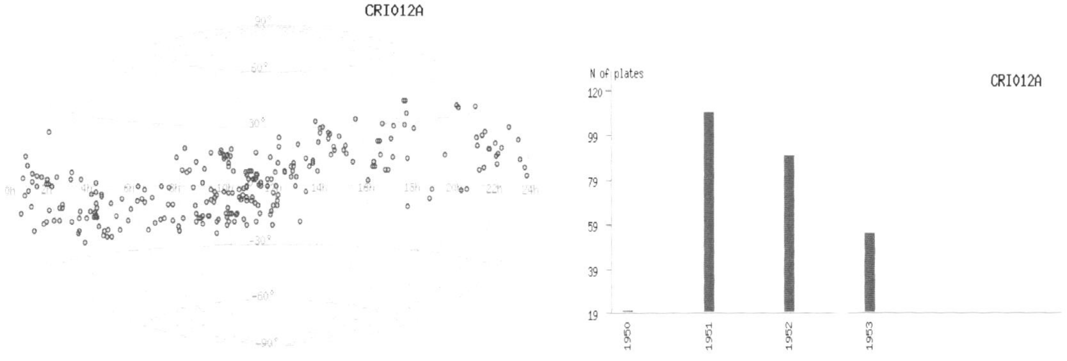

Figure 6: Plates distribution & Archive statistics CRI012A (1951 - 1953)
From DATABASE of JOINT PLATE ARCHIVE (DBGPA V2.0)
http://gua.db.ukr-vo.org/guides.php).

In the summer of 2014 to Odessa glass library, sponsors have donated the Epson Perfection V700 Photo scanner (the A4 format, optical resolution 6400x9600 dpi, depth of color, bit: 48) complete with office equipment serving the scanner. From this date digitization of the Odessa archive began to be conducted locally, not taking out of its far from its storage place.

The plates are digitized in 16-bit TIFF format, 20 μm / pixel. Size of one image is about 90 MB at 1200 dpi resolution and about 350 MB at 2400 dpi. Scanner Epson Perfection V700 was investigated by original methodology of V.

Andruk (Andruk, et al. 2015). As a result, he concluded that for rectangular coordinates mean-square error of value of one difference determination is σ_{xy} = 0.023 pixels for the 1200 dpi scanning mode; for instrumental stellar magnitudes, the mean-square error value of one difference determination is $\sigma_m = 0.013^m$ (Kashuba et al, 2017).

Figure 7: The workplace of the Epson Perfection V700 Photo Scanner operator.

At the preparatory stage of placing the digitized material in the database of the Bulgarian Data Processing Center (WFPDB) the following work was done:

- the original machine-readable catalog, recorded from observational logbooks and envelopes of the Simeiz collection and prepared in 2002 ((Pikhun, et al. 2002)) was presented in WFPDB format in 6 tables:

1. SIM012Aavi
2. SIM012Adigit
3. SIM012Amaindata
4. SIM012Anotes
5. SIM012Aobserver
6. SIM012Aquality

- the centers of the plates with use of service astrometry.net are specified;
- preview scans (See Figure 8, 9 for examples) with low resolution (600 dpi, jpg) were prepared;
- observational logbooks and part of the envelopes of the plates were digitized.

Figure 8: Preview pl. № 607=SIM012A000607.Brooks commet, 16.09.1911, exp. 150.0 min.

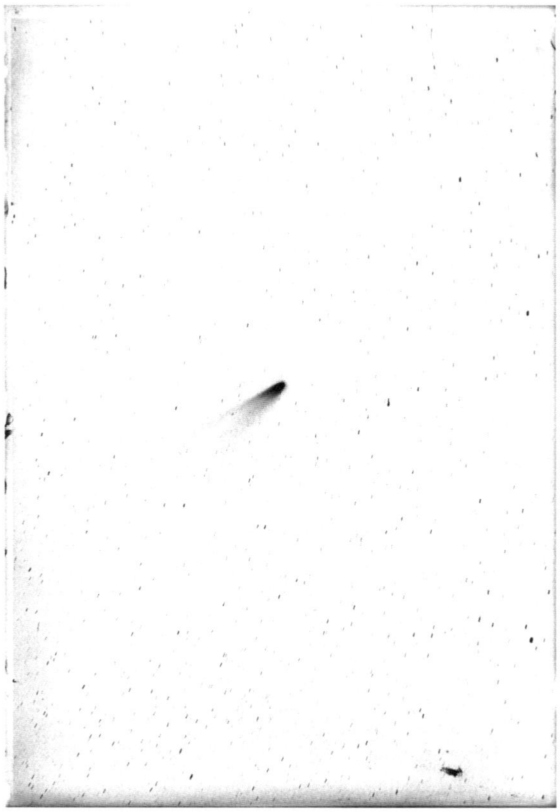

Figure 9: Preview pl. № 1464=SIM012A001464. Delavan commet, 20.09.1914, exp. 80.0 min.

In May 2018, the procesed data of Simeiz plates (887 SIM012A - 80GB) covered the period 1909-1915 were sent at Wide-Field Plate database in Sofia (See Fig. 10-12) (Kirov, et al. 2014).

Name	Last modified	Size	Description
Parent Directory		-	
SIM012A/	03-May-2018 15:20	-	
SIM012A_logs_1-2129/	08-May-2018 11:53	-	

Figure 10: Index of/ftp/WFPDB/archives/SIMEIZ.

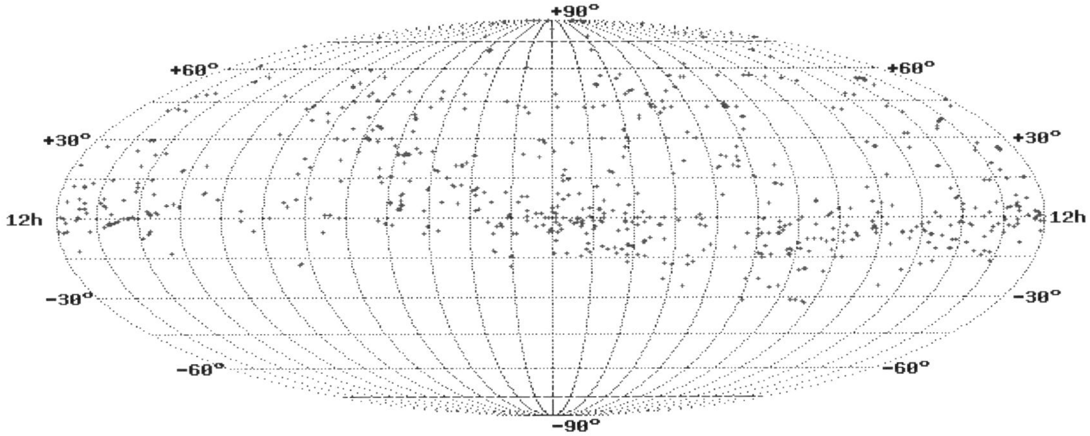

Figure 11: Distribution of 887 plates SIM012A on the sky in the Molveide projection.

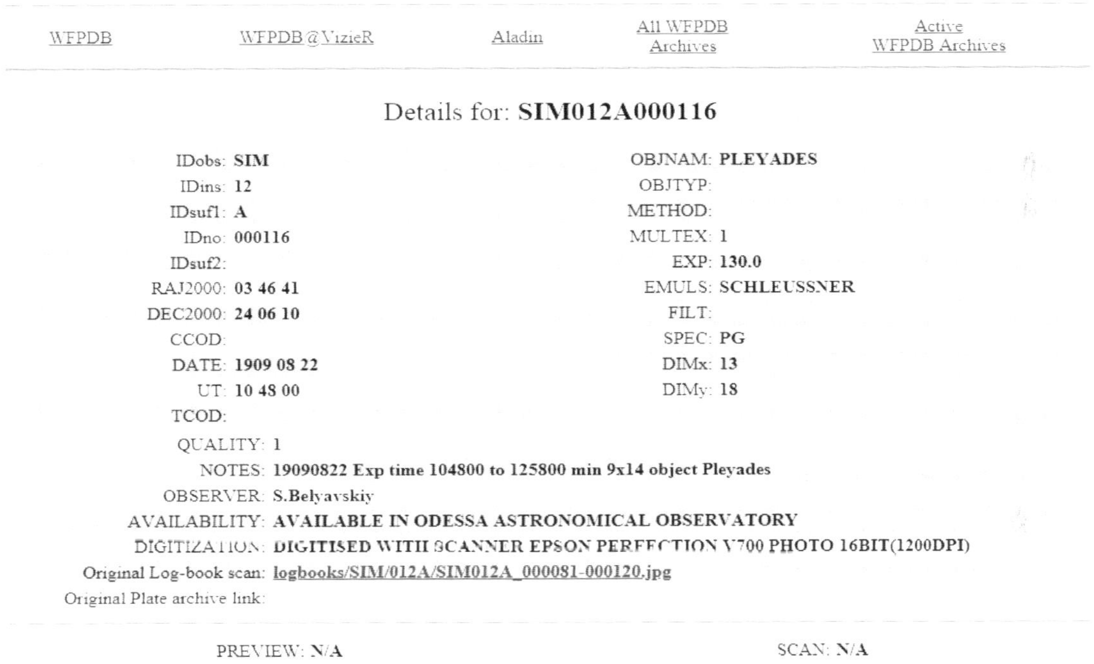

WFPDB | WFPDB@VizieR | Aladin | All WFPDB Archives | Active WFPDB Archives

Details for: **SIM012A000116**

IDobs:	**SIM**	OBJNAM:	**PLEYADES**
IDins:	**12**	OBJTYP:	
IDsuf1:	**A**	METHOD:	
IDno:	**000116**	MULTEX:	**1**
IDsuf2:		EXP:	**130.0**
RAJ2000:	**03 46 41**	EMULS:	**SCHLEUSSNER**
DEC2000:	**24 06 10**	FILT:	
CCOD:		SPEC:	**PG**
DATE:	**1909 08 22**	DIMx:	**13**
UT:	**10 48 00**	DIMy:	**18**
TCOD:			

QUALITY: **1**
NOTES: **19090822 Exp time 104800 to 125800 min 9x14 object Pleyades**
OBSERVER: **S.Belyavskiy**
AVAILABILITY: **AVAILABLE IN ODESSA ASTRONOMICAL OBSERVATORY**
DIGITIZATION: **DIGITISED WITH SCANNER EPSON PERFECTION V700 PHOTO 16BIT(1200DPI)**
Original Log-book scan: **logbooks/SIM/012A/SIM012A_000081-000120.jpg**
Original Plate archive link:

PREVIEW: **N/A** SCAN: **N/A**

Figure 12: Screenshot of the catalog page SIM012A in WFPDB (**www.wfpdb.org**).

References

Andrievsky, S., Pikhun, A., Kashuba, S., Tsvetkov, M., Tsvetkova, K.: 2007, "Odessa Astronomical Plate Collection", *Jenam*-2007, Yerevan.

Andruk, V. M., Pakuliak, L. K., Golovnia, V. V., Ivanov, G. A., Yizhakevich, O. M., Protsyuk, Yu. I., Shatokhina, S. V.: 2015, "Catalog of positions and b magnitudes of stars in the circumpolar region of northen sky survey (fon) project", *Odessa Astronomical Publications*, **28**, 192.

Karetnikov, V. G., Markina, A. K., Sotnikov, V. P.: 1994, "The Odessa Sky Patrol Plate Collection", Flares and Flashes; Proceedings of the IAU Colloquium No. 151; Held in

Sonneberg; Germany; 5 - 9 December 1994; XXII; 477 pp., Springer-Verlag Berlin Heidelberg New York. Also *Lecture Notes in Physics,* **454** Edited by Jochen Greiner, Hilmar W. Duerbeck, and Roald E. Gershberg, p. 407.

Karetnikov, V. G., Dorokhova, T. N., Yushchenko, A. V., Pikhun, A. I., Koshkin, N. I.: 2007, "The Digitization Project of Odessa Plate Depository", Library and Information Services in Astronomy V *ASP Conference Series*, **377.**

Kashuba, S., Andruk, V., Kashuba, V.: 2017, *Odessa Astronomical Publications*, **30**, 174.

Kirov, N., Tsvetkov, M., Tsvetkova, K.: 2014, "WFPDB: Software for Time and Coordinates Conversions", Proceedings of the IX Bulgarian-Serbian Astronomical Conference: Astroinformatics, (IX BSACA) Sofia, Bulgaria, July 2–4, 2014, Editors: M. K. Tsvetkov, M. S. Dimitrijević, O. Kounchev, D. Jevremović and K. Tsvetkova, Publ. Astron. Soc. "Rudjer Bošković", **15**, 2015, 43-48.

Pikhun, A.I., Yushchenko, A.V.: 2002, "Machine readable catalog of Odessa patrol plates"*, Information bulletin on variable stars* (*IBVS*), № **5215**.

Tsvetkov, M.: 2006, "Wide-Field Plate Database: a Decade of Development"*, Virtual Observatory: Plate Content Digitization, Archive Mining and Image Sequence Processing, iAstro workshop*, Sofia, Bulgaria, 2005, ISBN-10 954-580-190-5, p. 10.

Vavilova, I. B., Pakulyak, L. K., Shlyapnikov, A. A., Protsyuk, Yu. I., Savanevich, V. E., Andronov, I. L., Andruk, V. N., Kondrashova, N. N., Baklanov, A. V., Golovin, A. V., Fedorov, P. N., Akhmetov, V. S., Isak, I. I., Mazhaev, A. E., Golovnya, V. V., Virun, N. V., Zolotukhina, A. V., Kazantseva, L. V., Virnina, N. A., Breus, V. V., Kashuba, S. G., Chinarova, L. L., Kudashkina, L. S., Epishev, V. P.: 2012, "Astroinformation resource of the Ukrainian virtual observatory: Joint observational data archive, scientific tasks, and software", *Kinematics and Physics of Celestial Bodies*, **28(2)**, 85-102 (04/2012).

Proceedings of the XI Bulgarian-Serbian Astronomical Conference
(XI BSAC) Belogradchik, Bulgaria, May 14-18, 2018
Editors: Milcho K. Tsvetkov, Milan S. Dimitrijević and Momchil Dechev
Publ. Astron. Soc. "Rudjer Bošković" No 18, 2018, 217-223

ASYMMETRIC FILAMENT ERUPTION FOLLOWED BY TWO-RIBBON FLARE

KOSTADINKA KOLEVA, PETER DUCHLEV and MOMCHIL DECHEV

Institute of Astronomy with National Astronomical Observatory,
Bulgarian Academy of Sciences
E-mail: koleva@astro.bas.bg, pduchlev@astro.bas.bg, mdechev@astro.bas.bg

Abstract. We report the first results from the study of an asymmetric eruptive prominence (EP) that appeared on 01 Nov 2014 and was followed by a two-ribbon solar flare. The ejection triggered a fast coronal mass ejection (CME) that was well visible in the LASCO C2 field of view. The morphology and kinematics of the EP and two-ribbon flare were examined by multi-channel observations from AIA/SDO and SoHO/LASCO. Initially, the EP slowly rose and then it sharply ejected up with a strong acceleration producing the CME bright core. The evolution of two-ribbon flare is morphologically characterized by separation of the two ribbons in the chromosphere. The ribbons' separation showed two-stage evolution: first one with relatively fast decelerating motion and very slow second one with low constant velocity. Such separating motion is believed to provide a signature of the reconnection process occurring progressively higher up in the corona.

1. INTRODUCTION

Solar filaments (or prominence when observed on the limb) are relatively cool (10^4 K) and dense (10^{10}–10^{11} cm^{-3}) plasma objects embedded in the hot (10^6 K) and tenuous (10^9 cm^{-3}) solar corona (Tandberg-Hanssen 1995). Very often filaments displays eruptive motions. Solar prominence eruptions is one of the most energetic active phenomena of the Sun, during which about 10^{12} kg plasma can be thrown out from the chromosphere and low corona to the interplanetary space. Hence, the prominence eruptions has an important impact for the space weather manifestations. Filament eruptions are closely associated with CMEs.

Filaments are often observed to erupt asymmetrically. During the so called asymmetric eruption one of the prominence leg remains fixed to the photosphere, while the other undergo rising motions until its position becomes almost perpendicular to the solar limb (Liu et al. 2009).

Filament/prominence eruption are frequently accompanied by a two-ribbon solar flare, which are observed to expand (separate) in the chromosphere. It is widely accepted that two-ribbon flares are triggered by magnetic reconnection between coronal field lines that has been forced to open by erupting filaments (Svestka & Cliver 1992).

In this work we present the kinematic and morphological analysis of the asymmetric filament eruption that was followed by a two ribbon solar flare. The set of observations used for our analysis are described in Sect. 2. In Sect. 3 and Sect. 4 are presented the morphology and kinematics of the filament eruption and solar flare, respectively. In Sect. 5 some details of the associated CME are given. The main results and conclusions are summarized in Sect. 6.

2. OBSERVATIONS

The asymmetric prominence eruption was observed at 01 Nov 2014 in a quiet solar region close to the east limb. The eruption started at 04:26 UT and ended at 05:28 UT. The ejection of the prominence material was followed by a two ribbon solar flare, located at heliographic coordinates S22E52.

For event analyzed here we used observations from Atmospheric Imaging Assembly (AIA) on board the Solar Dynamics Observatory (SDO) (*Lemen, et al.* 2012). The AIA consists of seven Extreme Ultra-Violet field (EUV) and three Ultra-Violet (UV) channels, which provide an unprecedented view of the solar corona with an average cadence of ~12 s. The AIA image field-of-view reaches 1.3 solar radii with a spatial resolution of ~ 1.5". The eruption kinematics was studied by images taken with 1 min cadence in He II 304 Å passband. To trace the flare ribbon kinematics the data in Fe XIII 131Å, Fe IX 171 Å and C IV 1600 Å channels from SDO/AIA were also used.

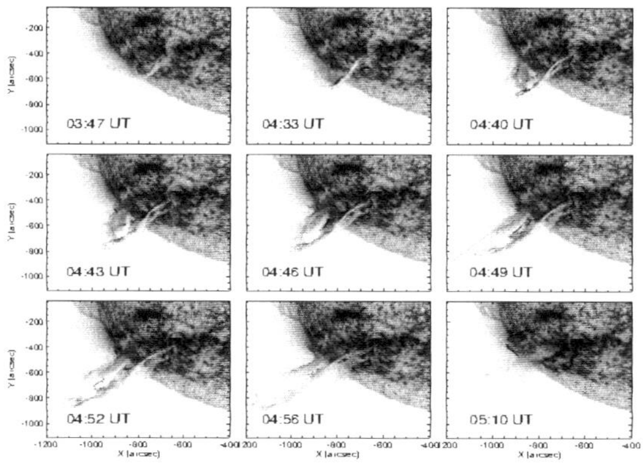

Figure 1: The EP evolution in AIA/SDO FOV in He II 304 Å channel (reversed colors). The green line marks the slice position used for obtaining the height-time evolution.

The ejection triggered a coronal mass ejection (CME). Images obtained by the Large Angle and Spectrometric Coronagraph (LASCO)/C2 on board SOHO, whose field-of-view extends from 2 to 6 solar radii (Brueckner et al. 1995) were also analyzed in order to trace the CME properties.

The evolution of prominence eruption in 304 channel of AIA/SDO in reversed colors is presented in Figure 1. With green line is marked the slice position used for kinematic study of the eruption.

3. PROMINENCE MORPHOLOGY AND KINEMATICS

The prominence body consisted of helically clockwise twisted flux robe (FR). The FR twist (a measure of the number of turns of the magnetic field lines) can be visually estimated after 04:43 UT at 3π (Fig. 1), which is above the critical value of 2π required for the kink instability to act (Hood & Priest 1981). At about 05:02 UT the left handed twist was transformed to the left-handed writhe. The same sign of the FR helical twisting and writhing is a basic condition for the kink instability presence (Hood & Priest 1981). That allow us to suggest that the eruptive prominence from 01 Nov 2014 was possibly subjected to a kink instability, although other instabilities may also have acted.

The prominence eruption started at about 03:47 UT. The eruption clearly showed two phases. During the first, slow-rise phase, which lasted at about 24 min the prominence body rose with velocities from 1.2 km/s to ~20 km/s. The second fast-rise phase lasted until 05:00 UT when the top of the prominence body quit the AIA field of view (FOV). During this phase the prominence ascended with increasing speed and acceleration. At the same time the prominence body underwent an untwisting motion. The fast rise phase velocity and acceleration reached 468 km/s and 700 m/s^2, respectively. After 05:09 UT the northern leg disconnected from the solar chromosphere.

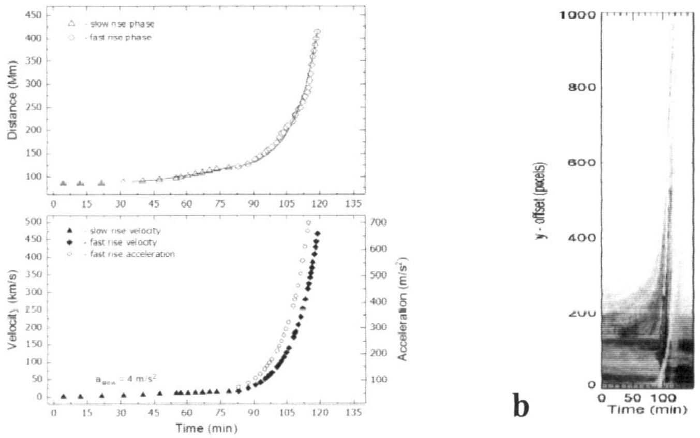

a b

Figure 2: Height-time profile of prominence evolution (top) and eruption velocities and acceleration (bottom) - (**a**). The start time is 01 Nov 2014 03:00 UT. Time-distance diagram – (**b**).

The height-time profile of prominence evolution and eruption velocities and acceleration are shown in Fig. 2a, while Fig 2b presents the time-distance diagram, used for height determination.

4. SOLAR TWO-RIBBON FLARE KINEMATICS

The prominence eruption form 01 Nov 2014 was followed by a C2.7 Goes class two-ribbon solar flare. The flare started at 04:44 UT, 12 minutes after the start of the EP fast-rise phase. It's peak intensity was registered at 05:43 UT. In Fig. 3 are presented the flare images in four AIA/SDO channels with different temperature formation, He II 304 Å (log(T)=4.7), Fe IX 171 Å (log(T)=5.8), C IV 1600 Å (log(T)=5.0) and Fe VIII 131 Å (log(T)=7.2). At such a way the flare evolution can be traced in different regions of solar atmosphere, such as chromosphere (304 Å), quiet corona and upper transition region (171 Å), upper photosphere (1600 Å) and hot flaring regions (131 Å).

It is widely accepted that the evolution of the two-ribbon solar flares is morphologically characterized by separation of the ribbons in the solar chromosphere. Such separating motion is believed to provide a signature of the reconnection process occurring in the corona (Cheng et al. 2003, Wang et al. 2003). In this paper we studied the separation of the flare ribbons observed at the chromosphere as a function of time. The ribbons' separation showed two-stage evolution. During the first stage which lasted about 43 minutes, the two ribbons moved away from each other with relatively fast decelerating velocity from 35 km/s to several km/s. The second stage was characterized by a motion with a low constant velocity of 1.4 km/s and lasted for more than 1.5 hr when flare emission decayed. The results are presented in Fig. 4.

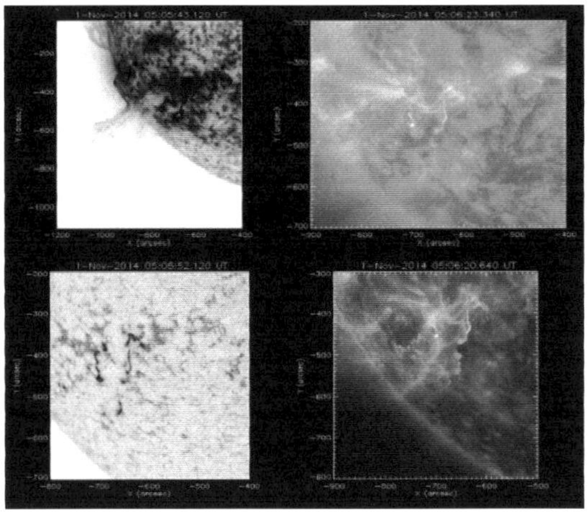

Figure 3: Two ribbons flare as viewed in AIA 304 Å, 171 Å, 1600 Å and 131 Å channels at temperature formation: log (T) = 4.7, 5.8, 5.0 and 7.2, respectively.

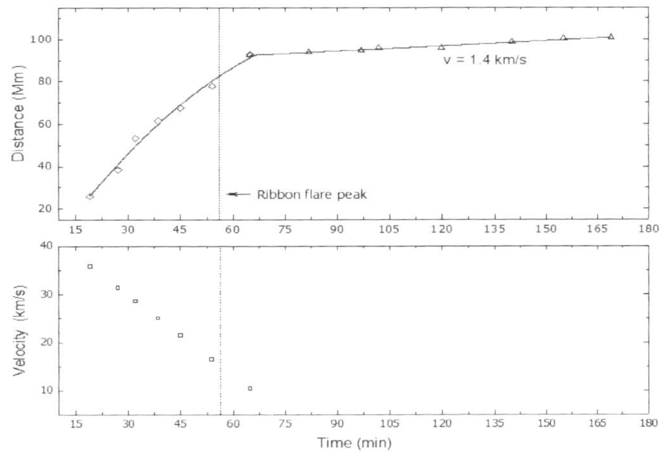

Figure 4: The distance between flare ribbons (top panel) and the separation velocity (bottom panel). The flare peak time is marked with dotted line.

5. THE ASSOCIATED CME

The prominence eruption was associated with a fast partial halo CME, which first appearance in LASCO/C2 FOV was at about 05:00 UT. The CME was located at position angle (PA) of 74° degrees and propagated with linear speed of 1624 km/s and acceleration of 7.3 m/s^2. The EP material could be traced in LASCO/C3 FOV up to 16 solar radii as the CME bright core. The CME evolution in LASCO C2 field of view is shown in Fig. 5, while in Fig. 6 the height-time profiles of EP and associated CME are presented together with Goes soft X-ray (SXR) flux in two energy channels.

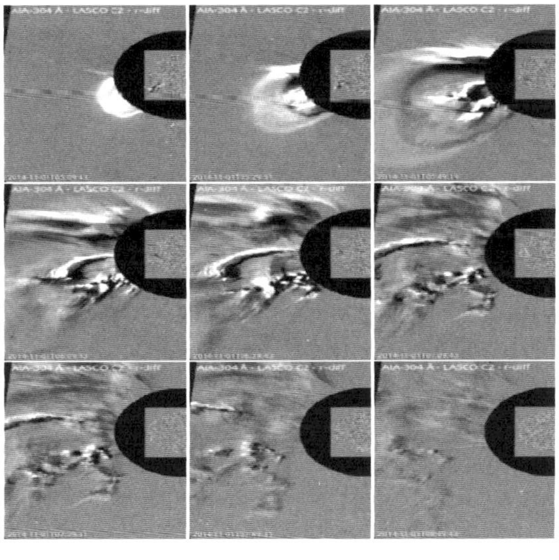

Figure 5: CME evolution sample in LASCO C2 field of view.

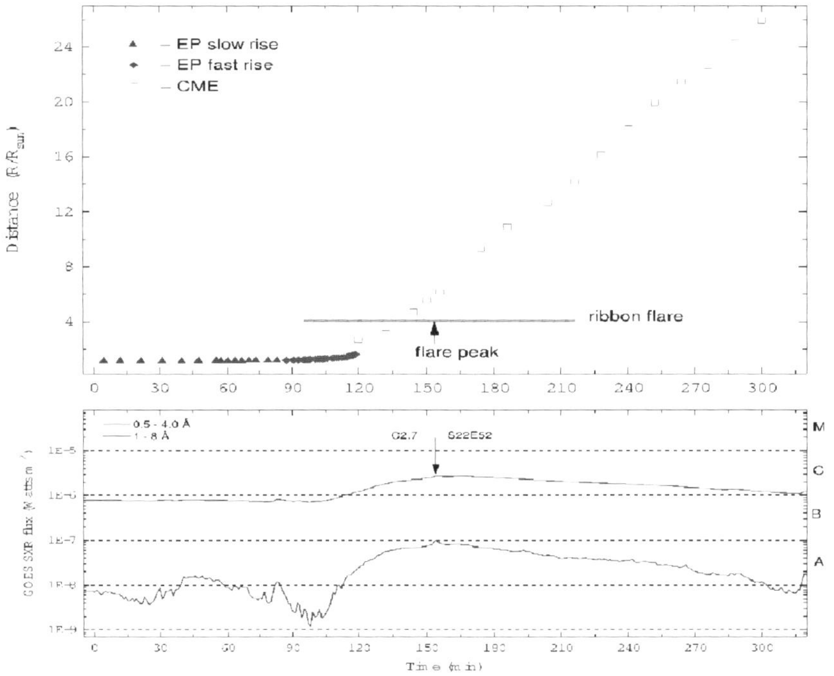

Figure 6: Height-time evolution of EP and associated CME (top panel) and Goes SXR flux (bottom panel). The start time is 03:00 UT. The time of flare peak is marked with black arrow.

6. SUMMARY

The prominence from 01 Nov 2014 underwent an asymmetric eruption with two clearly distinguish by its kinematic phases: slow-rise phase with a modest velocities and a phase of strong acceleration and very high speed. During the first phase the EP slowly rose with velocities from 1.2 km/s to ~18 km/s and then it underwent a strong acceleration (second phase) and rose with velocities from 20 km/s up to 468 km/s in the AIA/SDO field of view.

The two-ribbon flare underneath the EP most probably occurred due to reconnection processes in the coronal magnetic field in the wake of the prominence eruption. The ribbons' separation kinematics clearly shows two stages: first decelerating one with velocities from 35 km/s to several km/s and second very slow one with velocity of 1.4 km/s.

The EP produces a very fast halo CME, which propagates with velocity of 1624 km/s. The CME core formed by the EP can be traced up to 16 solar radii in the SOHO/LASCO C3 field of view.

Acknowledgment

This research was partially supported by the Bulgarian National Science Fund of the Ministry of Education and Science under grants DH 081/13.12.2016 and DN 18/13-12.12.2017.

References

Brueckner, G. E., Howard, R. A., Koomen, M. J., et al.: 1995, *Sol. Phys.*, **162**, 357.

Cheng, C. Z., Ren, Y., Choe, G. S., Moon, Y.-J.: 2003, *ApJ*, **596**,1341.

Hood, A. W., Priest, E. R.: 1981, *Geophys. Astrophys. Fluid Dyn.*, **17**, 297.

Lemen, J. R., Title, A. M., Akin, D. J., et al.: 2012, *Sol. Phys.*, **275**, 17.

Liu, R., Alexander, D., Gilbert, H.: 2009, *ApJ*, **691**, 1079.

Svestka, Z., Cliver, E. W.: 1992, in *Eruptive Solar Flares*, ed. Z. Svestka et al. (Berlin: Springer), 1.

Tandberg-Hanssen, E.: 1995, *The Nature of Solar Prominences*, D Reidel Publ. Co., Dordrecht.

Wang, H., Qiu, J., Jing, J., Zhang, H.: 2003, *ApJ*, **593**, 564.

Proceedings of the XI Bulgarian-Serbian Astronomical Conference
(XI BSAC) Belogradchik, Bulgaria, May 14-18, 2018
Editors: Milcho K. Tsvetkov, Milan S. Dimitrijević and Momchil Dechev
Publ. Astron. Soc. "Rudjer Bošković" No 18, 2019, 225-227

NEW VERSION OF SPECTRAVIEW FOR SPECTRA ANALYSIS

DRAGOMIR MARCHEV[1], GEORGI DIMITROV[2], NATALIYA PAVLOVA[2], DOROTEA VASILEVA[1], BORISLAV BORISOV[1] and TEODORA ATANASOVA[1]

[1] *Shumen University, Faculty of Natural Sciences, Department of Physics and Astronomy, 115 Universitetska Str., 9712, Shumen, Bulgaria*
E-mail: d.marchev@shu.bg, d.vasileva@shu.bg, b.borisov@shu.bg, t.atanasova@shu.bg
[2] *Shumen University, Faculty of Mathematics and Informatics, Department of Computer Informatics, 115 Universitetska Str., 9712, Shumen, Bulgaria*
E-mail: *g.dimitrov@shu.bg, n.pavlova@shu.bg*

Abstract: This article presents a new version of the program SpectraView (SV2). This version is equipped with new user interface and other features unavailable in its previous versions. It has also been created using modern programming language and is now compatible with the latest operating systems. The program has been tested on the observations of the four short-period variable stars.

1. INTRODUCTION

The article presents a new version of the program for 2D visualization of star spectra obtained in Rozhen NAO. The aim is to convert one-dimensional arrays into two-dimensional images with the possibility of adjusting the degree of gray. This allows us to visualize the curves of the radial velocities and to determine their half-amplitudes even more precisely. As known in astronomy, spectral observations are the only source of information on the masses of the stars. The program, object of the present work, aims to facilitate the analysis of the obtained radial velocity curves. The program has been tested on the observations of the stars NSVS 254037, TYC3621-711, ASAS J102556+2049,3 and NSVS 2569022.

2. DISCUSSION

The first two stars, that had been test objects of the initial version of the program, (Marchev and all, 2018), are presented with inverted graphics here – Fig.1 and Fig.2. This and the possibility to export the final images in different file formats are new functions of the program. Figures 3 and 4 show the converted images for the third and fourth stars respectively. Unfortunately due to the limits of the program version of MATLAB that we use, we are unable to make an EXE file version of the program. Source code and help file can be found at: http://astro.shu-bg.net/spectraview.zip . We are going to move to a newer version of the MATLAB program (R2016a), that allows us to achieve more functionality and better program interface. With the new version we are going to achieve much better visualization of the fitting procedures using the code of Snyder, L. F., 2001.

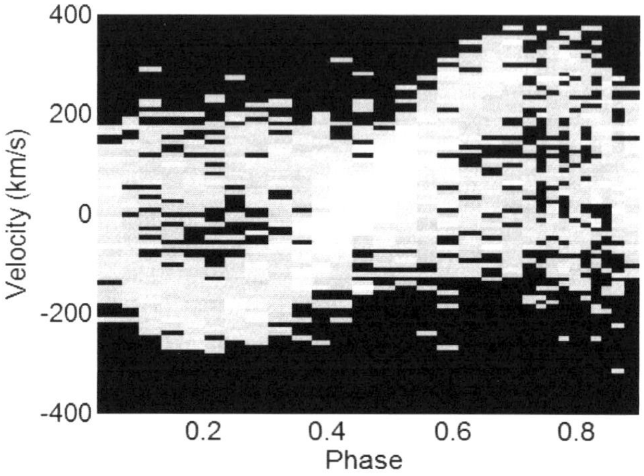

Figure 1: The star NSVS 254037 (negative conversion).

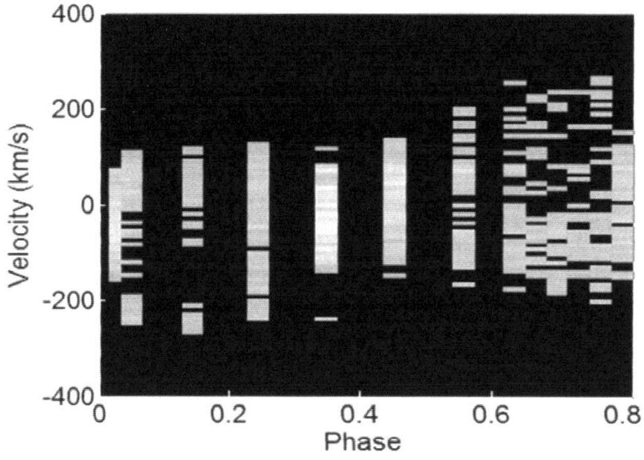

Figure 2: The star TYC3621-711.

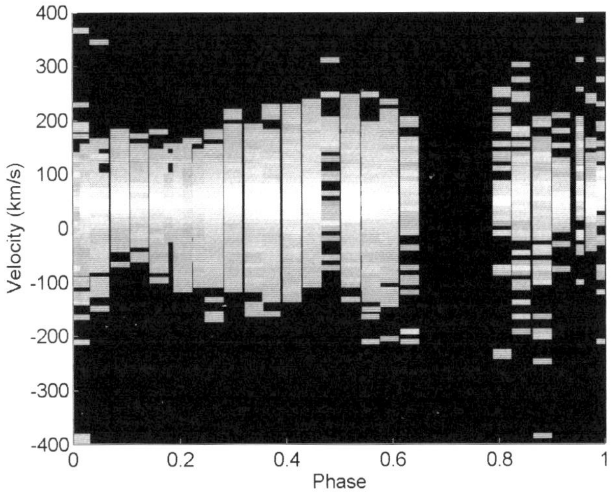

Figure 3: The star NSVS 2569022 (negative conversion).

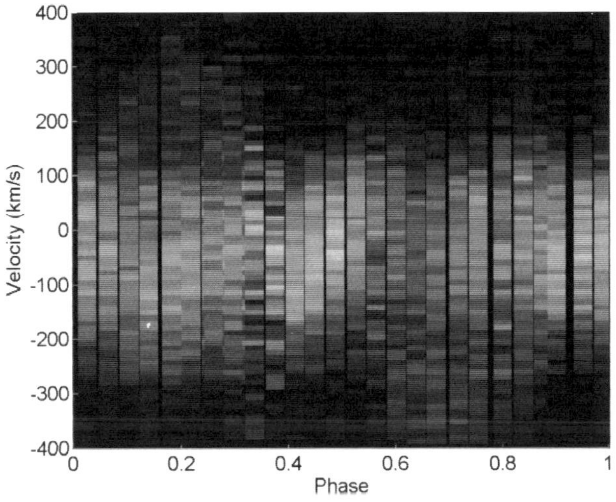

Figure 4: The star ASAS J102556.

Acknowledgements

The research was supported partly by project DN08/20 of Scientific Foundation of the Bulgarian Ministry of Education and Science as well as: RD-08-112/2018; RD-08– 164/2018 and RD-08-121/2018 of Shumen University.

References

Marchev, D.,V., and al.: 2018, "SpectraView 1.0 – 2D visualization of stellar spectra", *Acta Scientifica Naturalis*, **5(1)**, 35-39,

Snyder, L. F.: 2001, "Binary Maker 2.0 Light Curve Synthesis Program", *International Amateur-Professional Photoelectric Photometry Communication*, **83**, 25-33.

Proceedings of the XI Bulgarian-Serbian Astronomical Conference
(XI BSAC) Belogradchik, Bulgaria, May 14-18, 2018
Editors: Milcho K. Tsvetkov, Milan S. Dimitrijević and Momchil Dechev
Publ. Astron. Soc. "Rudjer Bošković" No 18, 2018, 229-237

PHOTOMETRIC STUDY OF UX ORI TYPE STARS GM CEP AND V1180 CAS

ASEN S. MUTAFOV[1], EVGENI H. SEMKOV[1], STOYANKA P. PENEVA[1] and SUNAY I. IBRYAMOV[2]

[1]Institute of Astronomy and National Astronomical Observatory, Bulgarian Academy of Sciences, 72, Tsarigradsko Shose Blvd., 1784 Sofia, Bulgaria
[2] Department of Physics and Astronomy, Faculty of Natural Sciences, University of Shumen, 115, Universitetska Str., 9712 Shumen, Bulgaria
E-mail: amutafov@astro.bas.bg, esemkov@astro.bas.bg, speneva@astro.bas.bg, sibryamov@shu.bg

Abstract. Results from BVRI long-term photometric observations of the pre-main sequence stars GM Cep and V1180 Cas are reported. The two stars were originally classified as eruptive variables from EX Lupi type. But the precise analysis of photometric data accumulated over the year's shows that they are certainly UX Orionis-type variables. Both types of pre-main sequence stars show variability with large amplitudes, but as in the case of EX Lupi stars variability is dominated of non-periodic outbursts, in the case of UX Orion stars long eclipses from circumstellar matter are observed. The analysis of the collected multicolor photometric data indicates the typical of UX Ori variables a color reversal during the minimums in brightness. This, along with the observed deep decreases in brightness, shows that stars are obscured by dust clumps or filaments passing through the line of sight to the stars. Photometric data presented in this study show the usefulness of systematically monitoring of pre-main sequence stars with large amplitude variability. With the accumulation of enough photometric data, we can more accurately determine the causes of photometric variability of young stellar objects.

INTRODUCTION

Photometric variability is a fundamental characteristic of the pre-main sequence (PMS) stars, which manifests as transient increases in brightness (outbursts), temporary drops in brightness (eclipses), irregular or regular variations for a short or long time scales. Both types of PMS stars the widespread low-mass (M =< 2 $M_{\odot}$) T Tauri Stars (TTSs) and the more massive Herbig Ae/Be (HAEBE) stars indicate photometric variability with various amplitudes and periods (Herbst et al. 1994; Herbst et al. 2007). A significant part of HAEBE stars and early type CTT stars show strong photometric variability with sudden quasi-Algol drops in brightness and amplitudes up to 2.5 mag. (V) (Natta et al. 1997; van den Ancker et al. 1998). During the deep minimums of brightness, an increase in polarization and specific color variability (called "blueing effect") are observed. The prototype of this group of PMS objects with intermediate mass named UXors is UX Orionis. The widely accepted explanation of its variability is a variable extinction from dust clumps or filaments passing through the line of sight to the star (Dullemond et al. 2003; Grinin et al. 1991).

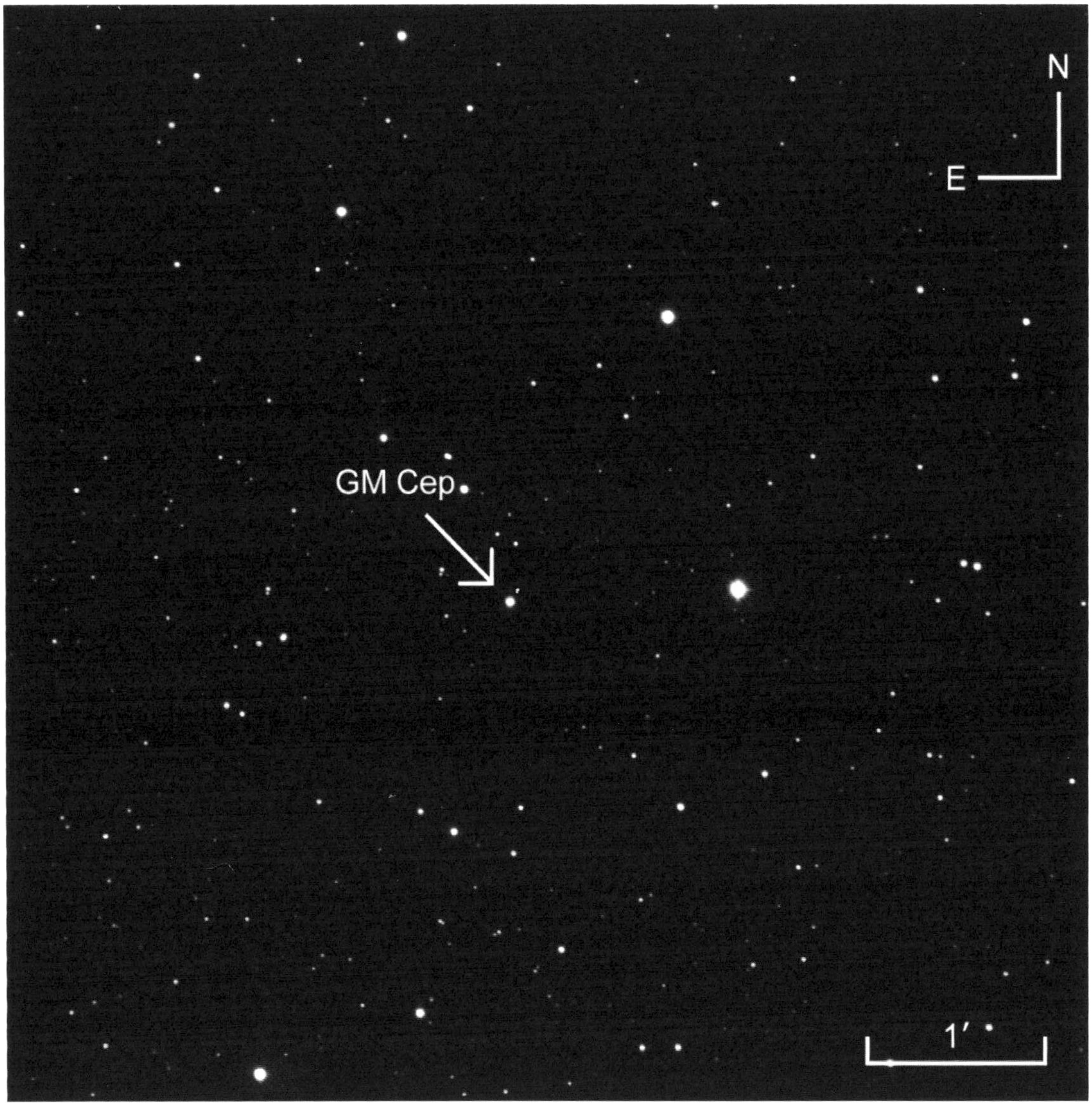

Figure 1: Color image of GM Cep obtained on November 23, 2016 with the 2-m RCC telescope in NAO Rozhen, Bulgaria.

Figure 2: Color image of V1180 Cas obtained on August 12, 2015 with the 1,3-m RC telescope in Skinakas Observatory, Greece.

The PMS star GM Cep lie in the field of the young open cluster Trumpler 37 (~4 Myr old) at a distance of 870 pc (Contreras et al. 2002) and most likely is a member of the cluster (Marschall & van Altena 1987; Sicilia-Aguilar et al. 2005). According to Sicilia-Aguilar et al. (2008) GM Cep is a PMS star with solar mass ($M \sim 2.1 M_{\odot}$) from G7V-K0V spectral type and with radius between 3 and 6 $R_{\odot}$. The results from *BVRI* optical photometric observations of the star collected in the period June 2008 – August 2014 are reported in our two previous papers (Semkov & Peneva 2012; Semkov et al. 2015). During our new photometric monitoring three deep minimums in brightness are observed. The collected multicolor photometric data shows the typical of UXor variables a color reversal during the minimums in brightness. Recent *BVRI* CCD photometric observations of GM Cep collected from November, 2014 to April, 2018 are reported here.

V 1180 Cas is a young variable associated with the dark cloud Lynds 1340 – a star forming region in Cassiopeia, located at a distance of 600 pc from the Sun. The star was first recognised to be a Há emitter by Kun et al. (1994). According to Kun et al. (2011) the object exhibited large amplitude (5–6 mag in the I_C band) brightness variations on 2–3 years timescales. Data from our *BVRI* photometric monitoring in the period October 2011 – April 2018 show drops in stellar brightness with amplitude up to 3 mag (I_c).

OBSERVATIONS

The CCD observation of GM Cep and V1180 Cas were performed in two observatories with four telescopes: 2-m Ritchey-Chretien-Coude, the 50/70-cm Schmidt and the 60-cm Cassagrain telescopes of the Rozhen National Astronomical Observatory (Bulgaria) and the 1.3-m Ritchey-Chretien telescope of the Skinakas Observatory of the University of Crete (Greece). The observations were performed with eight different types of CCD cameras. All frames were taken through a standart Johnson-Cousins set of filters. All data were analized using the same aperture, for GM Cep which was chosen to be with 6" radius (while the background annulus was taken from 11" to 17") and for V1180 Cas which was chosen to be with 4" radius (while the background annulus was taken from 13" to 17"). As a reference, the BVRI comparison for GM Cep sequence reported in Semkov & Peneva (2012) was used.

Figure 3: The used telescopes. Left to right: the 2-m RCC, the 50/70-cm Schmidt and the 60-cm Cassagrain telescopes of Rozhen NAO (Bulgaria) and the 1.3-m telescope of the Skinakas Observatory (Greece).

RESULTS

Long-term lights curves of GM Cep from all our observations and data from the literature are shown in Fig. 4. In Fig. 5 we plot the *B*, *V*, *R* and *I* light curves for the whole period of our CCD photometric monitoring (2008 - 2018). The typical values of instrumental errors for all of our observations are in the range 0.01^m-0.02^m (for *I* and *R*), 0.01^m-0.03^m (for *V*) and 0.01^m-0.05^m (for *B*). The new photometric data showed continued strong brightness variability of GM Cep as the registered in the previous studies (Sicilia-Aguilar et al. 2008; Xiao et al. 2010; Semkov & Peneva 2012; Chen et al. 2012, Semkov et al. 2015). Out of deep minimums GM Cep shows significant brightness variations in the time scale of days and months. The summarized results of over ten years period of observations show very strong photometric variability. We have registered eight deep minimum in brightness in the light curve of GM Cep.

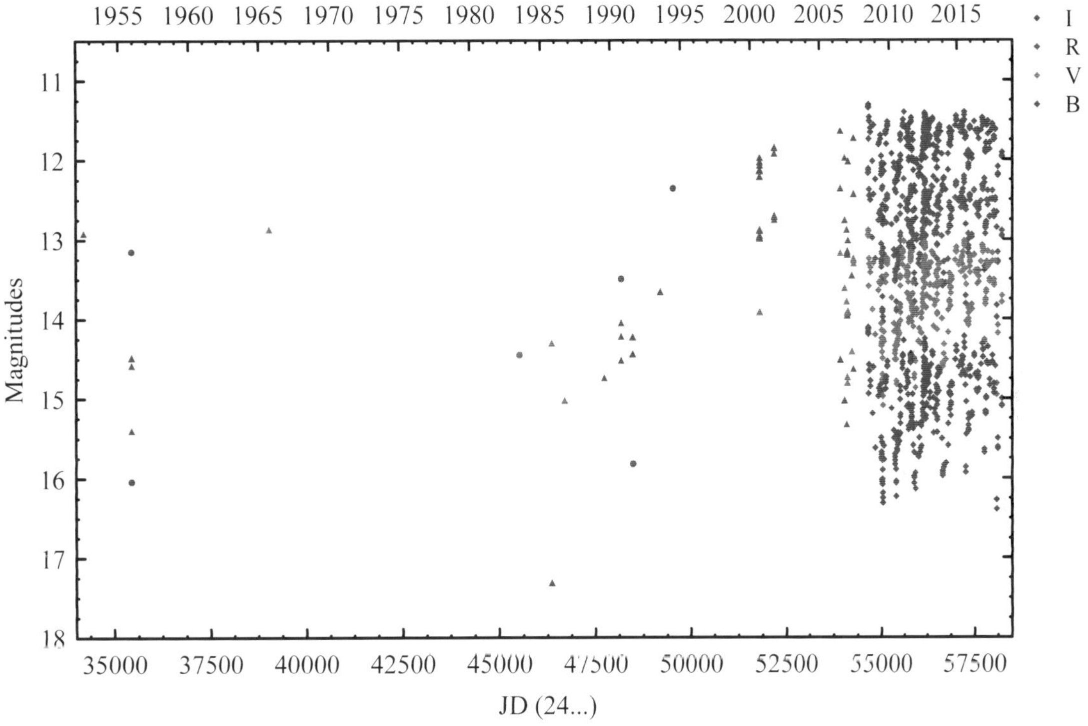

Figure 4: Long-term light curves of GM Cep. The diamonds denote our CCD photometric data; triangles – data from Sicilia-Aguilar (2008); circles – photographic data from POSS I and POSS II plates.

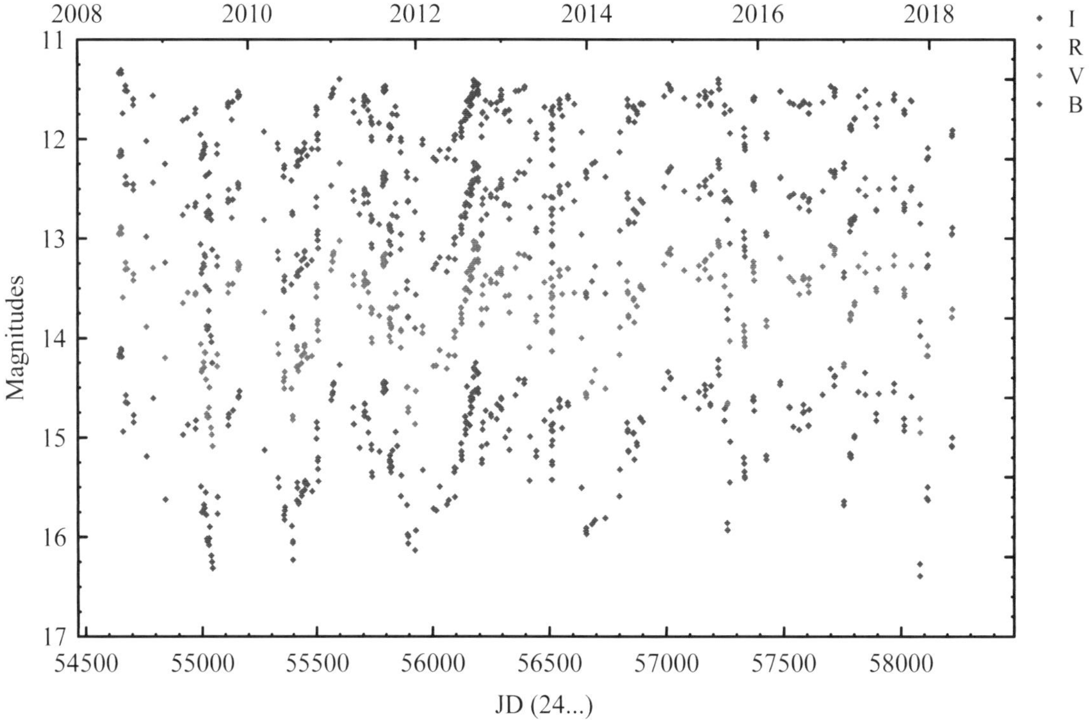

Figure 5: UBVRI light curves of GM Cep for the whole period of our photometric monitoring (2008 - 2018).

The collected multicolor photometric data shows the typical of UXor variables a color reversal during the minimums in brightness. Using data from our *BVRI* photometry the three color-magnitude diagrams (B - V/V , V - R/V and V - I/V) of the star are constructed and displayed on Fig. 6. The existence of a turning point of each of the diagrams is seen on the figure. After analysis of data collected our conclusion is that the photometric properties of GM Cep can be explained by superposition of both: (1) highly variable accretion from the circumstellar disk onto the stellar surface, and (2) occultation from circumstellar clumps of dust, planetesimals or from features of the circumstellar disk. Our photometric results for the period June 2008 – April 2018 suggest that the variable extinction dominates the variability of GM Cep.

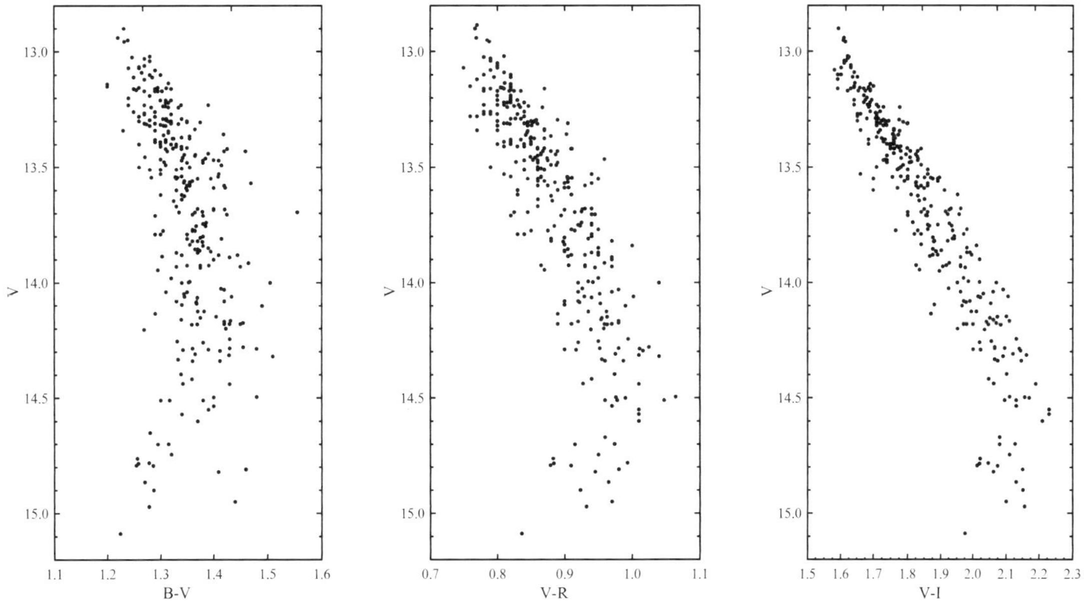

Figure 6: The color-magnitude diagrams of GM Cep in the period of observations June 2008 - April 2018.

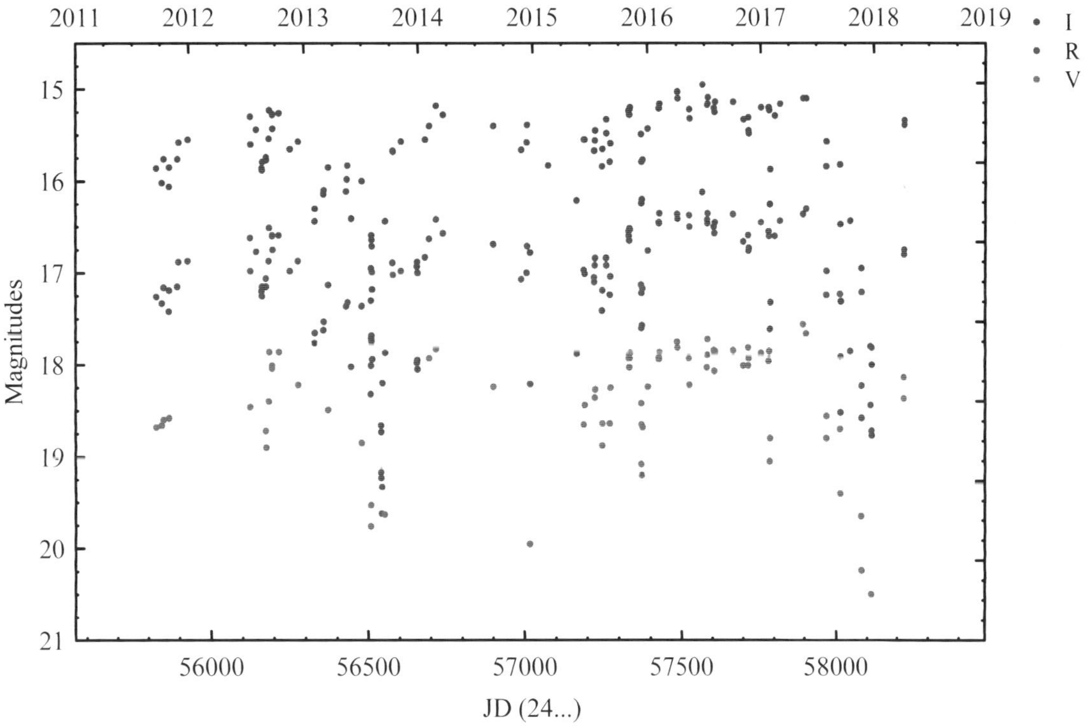

Figure 7: VRI light curves of V1180 Cas in the period October 2011 – April 2018.

Data from our *BVRI* photometric monitoring of V1180 Cas in the period October 2011 – April 2018 are presented in Fig.7. The photometric data show a large amplitude variations (ΔR ~3.3^{m}) during the 6.5 years period of observations. The first deep minimum is registered at September 2013 and the second at December 2017. Out of deep minimums V 1180 Cas shows significant brightness variations in the time scale of days and months, similar to another UX Ori type variable star GM Cep (Semkov & Peneva 2012; Semkov et al. 2015).

V 1180 Cas show color reversal at minimum of brightness a similar to GM Cep (Fig. 8).

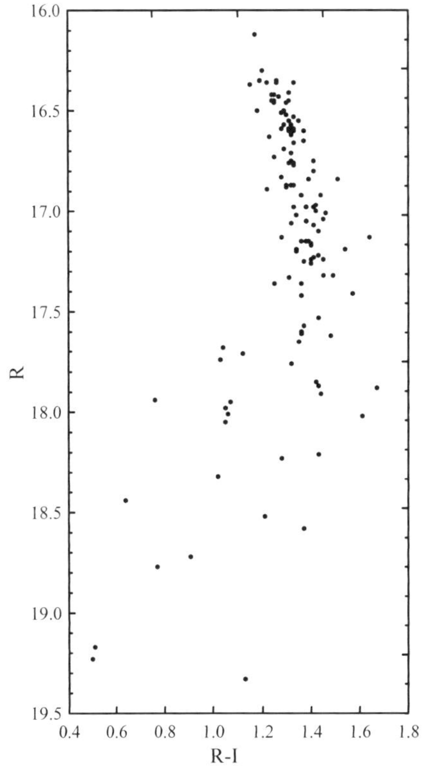

Figure 8: The color-magnitude diagrams of V1180 Cas in the period of observations October 2011 - April 2018.

Acknowledgements

This work was partly supported by the Bulgarian Scientific Research Fund of the Ministry of Education and Science under the grants DM 08-2/2016, DN 08-1/2016, DN 08-20/2016 and DN 18-13/2017 as well as by the project RD-08-112/2018 of the University of Shumen.

References

Chen, W. P., Hu, S. C. -L., Errmann, R. et al.: 2012, "A Possible Detection of Occultation by a Proto-planetary Clump in GM Cephei", *ApJ*, **751**, 118.

Contreras, M. E., Sicilia-Aguilar, A., Muzerolle, J., Calvet, N., Berlind, P., Hartmann, L.: 2002, "A Study of Intermediate-Mass Stars in Trumpler 37", *AJ*, **124**, 1585.

Dullemond, C. P., van den Ancker, M. E., Acke, B., van Boekel, R.: 2003, "Explaining UX Orionis Star Variability with Self-shadowed Disks", *ApJ*, **594**, L47.

Grinin, V. P., Kiselev, N. N., Minikulov, N. Kh., Chernova, C. P., Voshchinnikov, N. V.: 1991, "The investigations of 'zodiacal light' of isolated AE-Herbig stars with nonperiodic algol-type minima", ApSS, **186**, 283.

Herbst, W., Herbst, D. K., Grossman, E. J., Weinstein, D.: 1994, "Catalogue of UBVRI photometry of T Tauri stars and analysis of the causes of their variability", *AJ.*, **108**, 1906.

Herbst, W., Eisloffel, J., Mundt, R., Scholz, A.: 2007, "The Rotation of Young Low-Mass Stars and Brown Dwarfs" , in Protostars and Planets V, ed. B. Reipurth, D. Jewitt, K. Keil, 297-311.

Kun, M., Obayashi, A., Sato, F., Yonekura, Y., Fukui, Y., Balazs, L. G., Abraham, P., Szabados, L., Kelemen, J.: 1994, "Study of L 1340: A star-forming cloud in Cassiopeia", *A&A*, **292**, 249.

Kun, M., Szegedi-Elek, E., Moór, A., Kóspál, Á., Ábrahám, P., Apai, D., Kiss, Z. T., Klagyivik P., Magakian, T. Yu., Mező, Gy., Movsessian, T. A., Pál, A., Rácz, M., Rogers, J.: 2011, "A Peculiar Young Eruptive Star in the Dark Cloud Lynds 1340", *ApJ*, **733**, L8.

Marschall, L. A., van Altena, W. F.: 1987, "Membership in the young cluster Trumpler 37", *AJ*, **94**, 71.

Natta, A., Grinin, V. P., Mannings, V., Ungerechts, H.: 1997, "The Evolutionary Status of UX Orionis-Type Stars", *Astrophys. J.*, **491**, 885.

Semkov, E., Peneva, S.: 2012, "Optical photometry of GM Cep: evidence for UXor type of variability", *ApSS,* **338**, 95.

Semkov, E. H., Ibryamov, S. I., Peneva, S. P., Milanov, T. R., Stoyanov, K. A., Stateva, I. K., Kjurkchieva, D. P., Dimitrov, D. P., Radeva, V. S.: 2015, "The unusual photometric variability of the PMS star GM Cep", *PASA*, **32**, e011.

Sicilia-Aguilar, A., Hartmann, L., Hernández, J., Briceño, C., Calvet, N.: 2005, "Cepheus OB2: Disk Evolution and Accretion at 3-10 Myr", *AJ*, **130**, 188.

Sicilia-Aguilar, A., Merín, B., Felix, H., Ábrahám, P., Thomas, H., Mária, K., Nimesh, P., Attila, J., Wolfgang, B., Hartmann, Lee W., Csizmadia, Sz., Moor, A.: 2008, "The Rapid Outbursting Star GM Cep: An EXor in Tr 37?" *ApJ*, **673**, 382.

van den Ancker, M. E., de Winter, D., Tjin A Djie, H. R. E.: 1998, "HIPPARCOS photometry of Herbig Ae/Be stars",A&A, **330**, 145.

Xiao, L., Kroll, P., Henden, A.: 2010, Long-term Light Curve of Highly Variable Protostellar Star GM Cep, *AJ*, **139**, 1527.

Proceedings of the XI Bulgarian-Serbian Astronomical Conference
(XI BSAC) Belogradchik, Bulgaria, May 14-18, 2018
Editors: Milcho K. Tsvetkov, Milan S. Dimitrijević and Momchil Dechev
Publ. Astron. Soc. "Rudjer Bošković" No 18, 2018, 239-243

THE EFFECTS OF SOLAR ACTIVITY ON THE TERRESTRIAL LOWER IONOSPHERE

VLADIMIR A. SREĆKOVIĆ

Institute of Physics, University of Belgrade, P.O. Box 57, 11001 Belgrade, Serbia
E-mail: vlada@ipb.ac.rs

Abstract. The perturbations in the terrestrial D-region induced by solar flare X-ray energy were studied using monitored amplitude and phase data from Very Low Frequency (VLF, 3 - 30 kHz) and Low Frequency (LF, 30 - 300 kHz) radio waves. All data were recorded by Belgrade system of stations (44.85^0 N, 20.38^0 E). The focus of this contribution is on the study of perturbed amplitude on VLF/LF signal caused by strong solar flares. The results show that the magnitude of the VLF perturbations is in correlation with intensity of X-ray. The model computation and simulation are applied to acquire the electron density enhancement induced by intense solar radiation.

1. INTRODUCTION

The plasma in the ionospheric D-region (50 km $\leq$ h $\leq$ 90 km) is a very sensitive medium to external perturbation like moderate solar influence, stellar explosive radiation, etc. (Šulić & Srećković (2014) and Nina et al. (2011)). Processes like solar emission in far-UV and extreme-UV regions (Srećković et al. (2014); Mihajlov et al. (2013)) strongly affect the Earth's atmosphere. This intense solar radiation and activity can cause sudden ionospheric disturbances (SIDs) and further create ground telecommunication interferences, blackouts and minor radiation storms (see Radovanović et al. (2017)).

Solar flares that can be detected with VLF/LF radios waves are generally caused by X-ray flares and have various flux levels associated with them. Lower ionosphere monitoring by the mean of the radio technique can play a important role for a better understanding of space weather conditions Mitra (1974).

2. RESULTS

In this paper we confine our attention to the analysis of the amplitude and phase data of Very Low Frequency (VLF, 3 - 30 kHz) and Low Frequency (LF, 30 - 300 kHz) radio signals emitted by worldwide transmitters during SIDs.

The data were recorded at the Belgrade site with two receiver systems: Absolute Phase and Amplitude Logger (AbsPAL) system (Šulić et al. (2016)) and Atmospheric Weather Electromagnetic System for Observation Modeling and Education (AWESOME) (Šulić & Srećković (2014)). These data are digitized and saved in two different resolutions - high resolution (50 Hz) and low resolution (1 Hz). The analysis and comparison of VLF data has been carried out together with the examination of the corresponding solar X-ray fluxes. The intensity of the solar X-ray flux is taken from the online database of the GOES satellites.

2.1. Signal Monitoring

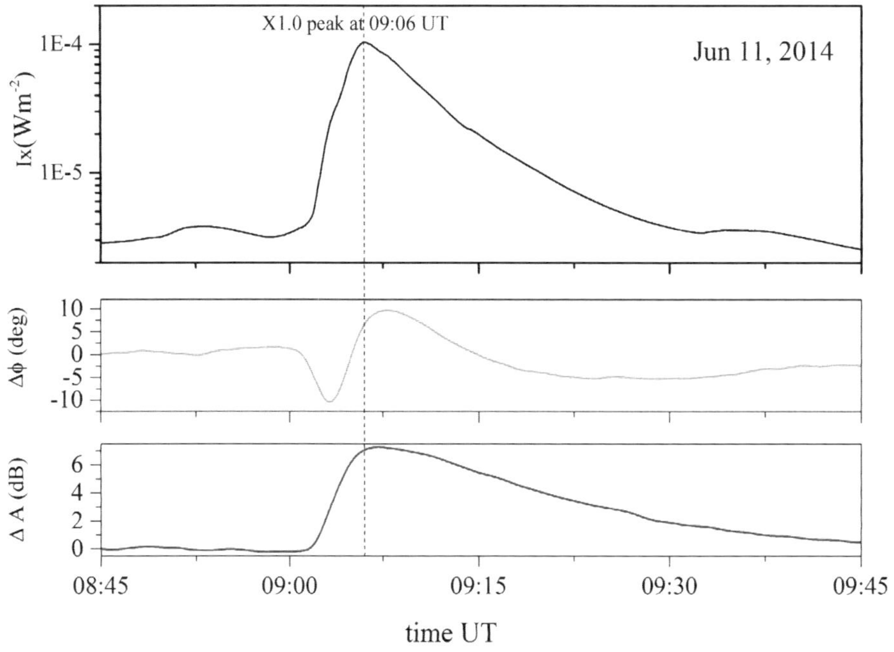

Figure 1: Simultaneous variation of the X-ray flux, phase and amplitude in the GQD/22.10 kHz radio signal (in universal time) recorded at Belgrade on 11 Jun 2014.

Simultaneous observations of amplitude (A) and phase (ϕ) in VLF/LF signals during solar flares could be applied for investigations of the perturbed ionospheric plasma. Therefore, the perturbation of the amplitude was estimated as a difference between values of the amplitude induced by some disturbance and amplitude in the normal ionospheric condition: $\Delta A = A_{per} - A_{nor}$. In subscript 'per' means the perturbed and 'nor' means normal condition. Similarly, the perturbation of phase was estimated as $\Delta\phi = \phi_{per}$-ϕ_{nor}.

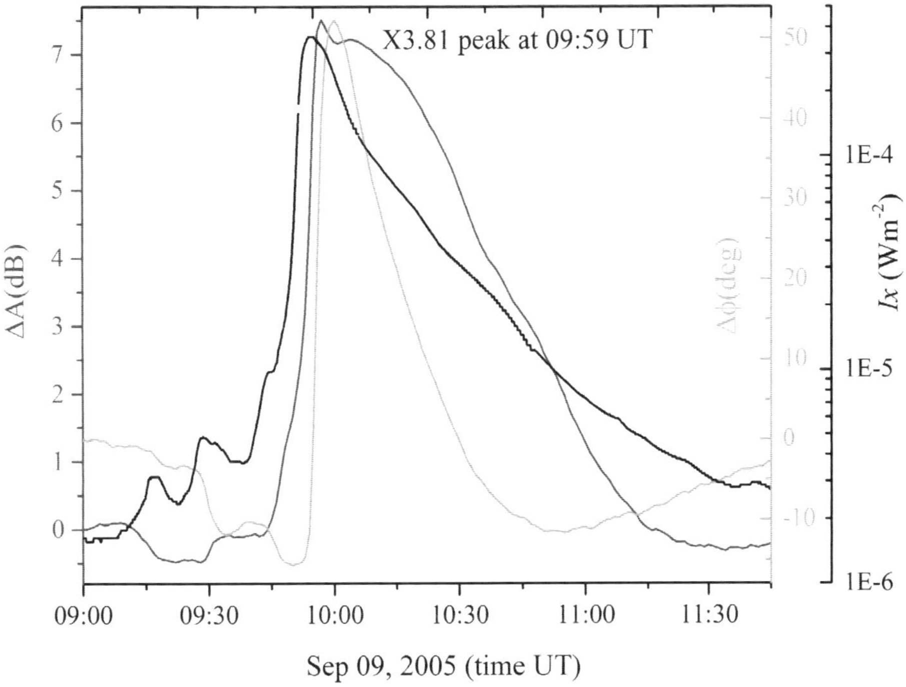

Figure 2: Simultaneous variation of the X-ray flux, phase and amplitude in the GQD/22.10 kHz radio signal (in universal time) recorded at Belgrade on 09 September 2005.

During the occurrence of solar flares, classified as a minor and small flare up to the C3 class, the amplitude of the VLF/LF signals does not have significant changes. A solar flare in the range from C3 to M3 classes induced an increase of the amplitude, which corresponds nearly proportional to the logarithm of the X-ray irradiance maximum (Šilić & Srećković 2014, Srećković et al. 2017).

2.2. Perturbations on VLF/LF Signal Induced by Strong Solar Flares

For studding SID we have selected strong solar flare event whose occurrence was in time interval of few hours around local noon at Belgrade. The selected events are X1.0 class solar flare recorded on 11 Jun 2014 and X3.81 class solar flare recorded on 09 September 2005. Our data are presented in Figs.1 and 2. Perturbations of the GQD/22.10 kHz radio signal are presented as temporal changes of ΔA and $\Delta \phi$ during solar flare event. Fig 3. presents calculated electron density at height h = 74 km during flare occurrences, as a function of the maximal intensity of X-ray flux. Electron density is calculated (as in Srećković et al. 2017) on the basis of VLF/LF propagation data recorded at Belgrade for one year. From the figures we conclude that changes of the phase and amplitude of radio signals during X class solar flares exhibit as well-defined enhancements that follow the development of the maximum in X-ray radiation. Also, after occurrences of such

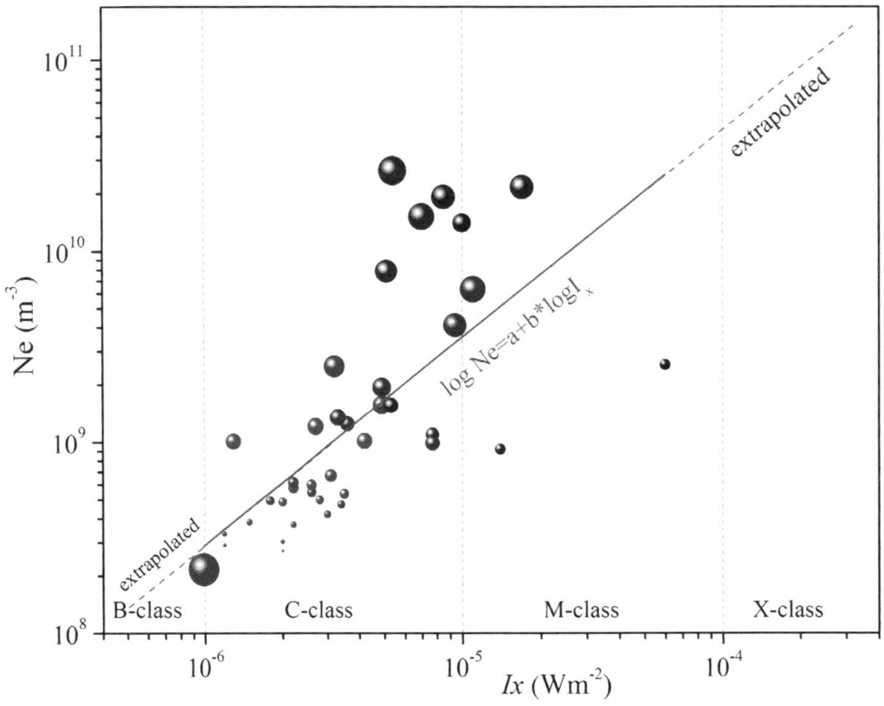

Figure 3: Values of electron density at height h = 74 km during flare occurrences, against maximum intensity of X-ray flux. Electron density is calculated on the basis of VLF/LF propagation data recorded at Belgrade for one year. The size of the point is proportional to the amplitude change while the color depends on the phase change (the darker color indicates a larger phase change). The red line indicates linear fit and red dashed line extrapolation to stronger/weaker solar flares.

strong solar flares the amplitude recovers bit slower than the phase. Using linear fits and extrapolations we can estimate the values of electron density even for larger solar X-ray flares (right dashed area) on the base of statistical analysis.

2.3. Perturbations on VLF/LF Signal Induced by Weak Solar Flares

Using linear fits and extrapolations we can estimate also the values of electron density even for week solar X-ray flares i.e. B-class on the base of statistical analysis. In Fig 3. is presented electron density at height h = 74 km during flare occurrences, as a function of maximal intensity of X-ray flux. The red line indicates linear fit and left red dashed line extrapolation to weaker solar flares.

3. CONCLUSIONS

In this work the effects during the enhancements of X-ray flux due to solar flares, on the propagating radio signal have been studied. The presented examples (Figs. 1, 2 and 3) are used to qualitatively describe consequences of SIDs, during occurrences of strong and weak solar flares.

Finally, we can point out the most important conclusion of this study: the successful use of applied technique for detecting space weather phenomena such as solar explosive events as well as for describing and further modeling the ionospheric plasma which is important as the part of electric terrestrial-conductor environment.

Acknowledgments

This work is partially supported by Ministry of Education, Science and Technological Development of the Republic of Serbia under the grants III 44002, and 176002.

References

Mihajlov, A. A., Ignjatović, L. M., Srećković, V. A., Dimitrijević, M. S., Metropoulos, A.: 2013, „The non-symmetric ion–atom radiative processes in the stellar atmospheres", *Mon. Notices Royal Astron. Soc.*, **431**, 589 – 599.

Mitra, A. P.: 1974, *Ionospheric Effects of Solar Flares*. eds. D.Reidel, Holland.

Nina, A., Čadež, V. M., Srećković, V. A., Šulić, D.: 2011, „The influence of solar spectral lines on electron concentration in terrestrial ionosphere", *Open. Astron.*, **20**, 609- 612.

Radovanović, M., Gomes, J. F. P., Yamashkin, A. A., Milenković, M., Stevančević, M.: 2017, „Electrons or protons: what is the cause of forest fires in western Europe on June 18, 2017", *J. Geogr. Inst. Cvijic, SASA*, **67(2)**, 213.

Srećković, V. A., Mihajlov, A. A., Ignjatović, Lj. M., Dimitrijević, M. S.: 2014, „Ion-atom radiative processes in the solar atmosphere: quiet Sun and sunspots", *Adv. Space Res.*, **54**, 1264-1271.

Srećković, V. A., Šulić, D. M., Vujčić, V., Jevremović, D., Vyklyuk, Y.: 2017, „The effects of solar activity: electrons in the terrestrial lower ionosphere", *J. Geogr Inst. Cvijic, SASA*, **67**, 3.

Šulić, D. M., Srećković, V. A: 2014, „A comparative study of measured amplitude and phase perturbations of VLF and LF radio signals induced by solar flares", *Serb. Astron. J.*, **188**, 45- 54.

Šulić, D. M., Srećković, V. A., Mihajlov A. A.: 2016, „A study of VLF signals variations associated with the changes of ionization level in the D-region in consequence of solar conditions", *Adv. Space Res.*, **57**, 1029-1043.

Proceedings of the XI Bulgarian-Serbian Astronomical Conference
(XI BSAC) Belogradchik, Bulgaria, May 14-18, 2018
Editors: Milcho K. Tsvetkov, Milan S. Dimitrijević and Momchil Dechev
Publ. Astron. Soc. "Rudjer Bošković" No 18, 2018, 245-248

THE RATE COEFFICIENTS OF THE SLOW ATOM-RYDBERG ATOM COLLISIONS IN GEOCOSMICAL PLASMAS

VLADIMIR A. SREĆKOVIĆ[1], MILAN S. DIMITRIJEVIĆ[2] and LJUBINKO M. IGNJATOVIĆ[1]

[1]*Institute of Physics, University of Belgrade, P.O. Box 57, 11001 Belgrade, Serbia*
[2] *Astronomical Observatory, Volgina 7, 11060 Belgrade, Serbia*
[3]*LERMA, Observatoire de Paris, PSL Research University, CNRS, Sorbonne Universités, UPMC (Univ. Pierre \& Marie Curie) Paris 06, 5 Place Jules Janssen, 92190 Meudon, France*
E-mail: vlada@ipb.ac.rs; mdimitrijevic@aob.bg.ac.rs; ljuba@ipb.ac.rs

Abstract. The processes in slow-atom-Rydberg-atom collisions are investigated in this contribution. The calculation of rate coefficients is done for Li and Na for the principal quantum numbers n≥4 and temperatures T≥500K. The chemi-ionization processes in slow collisions of excited alkali atoms with atoms in ground and excited states were considered, with a particular accent to the applications to geocosmical plasmas. We intend to present the results of calculation of the rate coefficients of the corresponding chemi-ionization processes in the tabulated form easy for further use.

1. INTRODUCTION

Finding new calculation possibilities and experimental techniques, has been continually inciting interest in the chemi-ionization processes in atom–Rydberg atom collisions, which resulted in numerous experimental/theoretical papers dedicated to this thematic (see e.g. Ignjatović et al. (2005)). The main aim of this work is to extend the investigation of Li and Na collisional ionization presented in Ignjatović et al. (2008) for the wider region of principal quantum numbers and temperature with a particular accent to the applications for low temperature laboratory Yubero et al. (2007) and astrophysical plasma research Klyucharev et al (2007) and Gnedin et al. (2009). We studied the non-symmetric chemi-ionization processes

$$X^*(n) + Y \Rightarrow e + XY^+ \qquad (1)$$

$$X^*(n) + Y \Rightarrow e + X^+ + Y \qquad (2)$$

where Rydberg atom X =Li and Y =Na, LiNa+ is the molecular ion in the electronic ground state ($X^2\Sigma^+$). The resonant mechanism method is used for the calculations of the rate coefficients of the processes (1) and (2). The description and basic theory is presented in the papers Ignjatović et al. (2005), Mihajlov et al. (2012). Here, the calculations of these rate coefficients are performed for the principal quantum number $4 \leq n \leq 25$ and temperatures $500\ \mathrm{K} \leq T \leq 1500\ \mathrm{K}$. The partial and the total rate coefficients which characterize the processes (1) and (2) separately and together can be presented by $K_{ci}^{(a)}(n;\ T)$, $K_{ci}^{(b)}(n;\ T)$ and $K_{ci}^{(ab)}(n;\ T)$, where

$$K_{ci}^{(ab)}(n,T) = K_{ci}^{(a)}(n,T) + K_{ci}^{(b)}(n,T). \qquad (3)$$

By definition, rate coefficients K(ab)ci (n; T) and K(a)ci (n; T) and the cross-sections $\sigma_{ci}^{(a,ab)}(n;\ T)$ are calculated as in Ignjatović et al. (2005) or Mihajlov et al. (2012,2017).

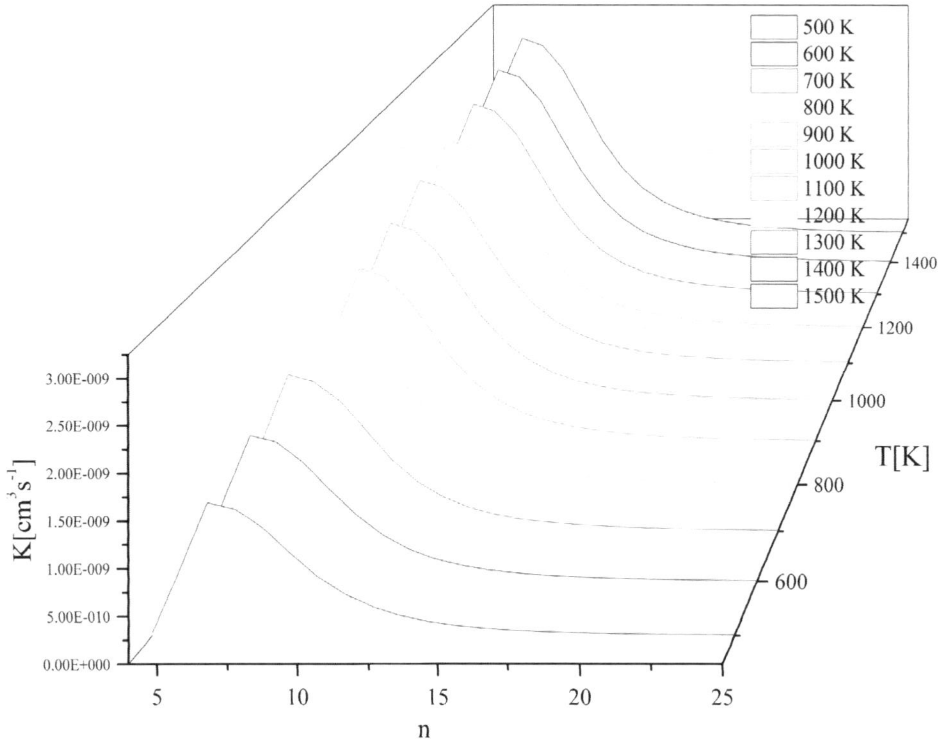

Figure 1: The total rate coefficient $K_{ci}^{(ab)}(n;\ T)$, Eq. (4) for chemi-ionization processes (1) and (2) in Li*(n) + Na collisions.

2. RESULTS AND DISCUSSION

We calculate the partial $K_{ci}^{(a)}(n;\ T)$, $K_{ci}^{(b)}(n;\ T)$ and total chemi-ionization rate coefficients $K_{ci}^{(ab)}(n;\ T)$, for the extended regions of principal quantum numbers n≥4 and temperatures $500\ \mathrm{K} \leq T \leq 1500\ \mathrm{K}$. Fig.1 present the plot of total chemi-ionization rate coefficient $K_{ci}^{(ab)}(n;\ T)$ for all mentioned conditions. One can see that the rate coefficient $K_{ci}^{(ab)}(n;\ T)$ has the maximum at n=6. For lower quantum numbers n the rate coefficients strongly depend on the temperature.

The relative contribution of the associative channel (1), is presented by Fig 2. It can be noticed that in the considered regions of n and T the associative chemi-ionization (1) dominates in comparison with the non-associative chemi-ionization channel (2).

Unlike the symmetric case, in the non-symmetric one associative channel changes non-monotonically with a maximum displaced to the region of small values of n.

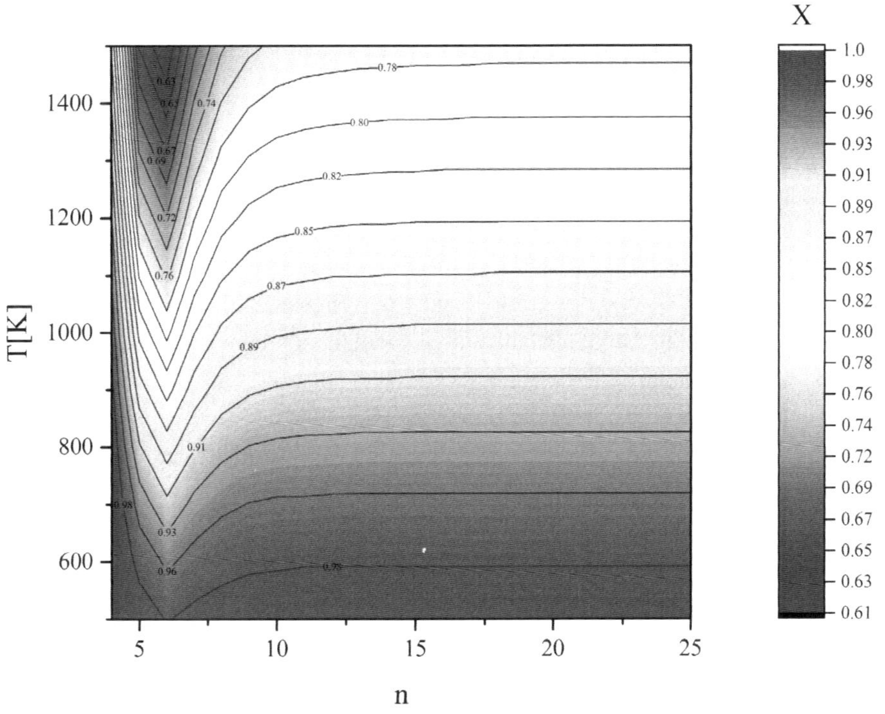

Figure 2: The ratio $O^{(ab)}$ of the partial rate coefficient $K_{ci}^{(a)}(n;\ T)$ and total ones $K_{ci}^{(ab)}(n;\ T)$ for chemi-ionization processes (1) and (2).

3. CONCLUSIONS AND PERSPECTIVES

The rate coefficients for the chemi-ionization processes in $Li^*(n)$ + Na collisions were calculated. The results of this work confirm the possibility of application of the resonant mechanism O'Keeffe et al. (2012) for the description of chemi-ionization collision processes, not only in the case, when the particles of the same type collide (symmetric case), but for the case of different type (non-symmetric) too. In the near future we aim to further investigate the chemi-ionization processes and develop completely quantum-mechanical methods of their description which could be applied in the cases of extremely low temperatures. Results of this work provide part of kinetics for the modelling of sodium formations in Io atmosphere, not investigated before. The other needed processes are well known and could be easily included in the model.

Acknowledgments

This work is partially supported by Ministry of Education, Science and Technological Development of the Republic of Serbia under the grants III 44002, and 176002.

References

Ignjatović, Lj. M., Mihajlov, A. A., Klyucharev, A. N.: 2008, „The rate coefficients of the chemi-ionization processes in slow Li*(n)+ Na collisions", *J. Phys. B* **41, 2**, 025203.

Yubero, C., Dimitrijević, M. S., Garcia, M. C., Calzada, M. D.: 2007, „Using the van der Waals broadening of the spectral atomic lines to measure the gas temperature of an argon microwave plasma at atmospheric pressure", *Spectrochim. Acta B*, **62, 2,** 169-176.

Klyucharev, A. N., Bezuglov, N. N., Matveev, A. A., Mihajlov, A. A., Ignjatović, Lj. M., Dimitrijević, M. S.: 2007, "Rate coefficients for the chemi-ionization processes in sodium-and other alkali-metal geocosmical plasmas", *New Astron. Rev.*, **51, 7**, 547-562.

Gnedin, Y. N., Mihajlov, A. A., Ignjatović, Lj. M., Sakan, N. M., Srećković, V. A., Zakharov, M. Y., Bezuglov, N. N., Klycharev, A. N.: 2009, „Rydberg atoms in astrophysics", *New Astron. Rev.* **53, 7-10,** 259-265.

Ignjatović, Lj. M., Mihajlov, A. A.: 2005, „Rate coefficient for the chemi-ionization in slow Li*(n)+ Li and Na*(n)+ Na collisions", *Phys. Rev. A*, **72, 2**, 022715

Mihajlov, A. A., Srećković, V. A., Ignjatović, Lj. M., Klyucharev, A. N.: 2012, „The chemi-ionization processes in slow collisions of Rydberg atoms with ground state atoms: mechanism and applications", *J. Clust. Sci.*, **23, 1,** 47-75

Mihajlov, A. A., Srećković, V. A., Ignjatović, Lj. M., Simić, Z., Dimitrijević, M. S.: 2017, „Influence of Rydberg atom-atom collisional and (n-n')-mixing processes on optical properties of astrophysical and low-temperature laboratory plasmas", *J. Phys. Conf. Ser.*, **810, 1**, 012058

O'Keeffe, P., Bolognesi, P., Avaldi, L., Moise, A., Richter, R., Mihajlov, A. A., Srećković, V. A., Ignjatović, Lj. M.: 2012, „Experimental and theoretical study of the chemi-ionization in thermal collisions of Ne Rydberg atoms", *Phys. Rev. A*, **85, 5**, 052705

Authors' Index

AUTHORS' INDEX

A

B

C

D

G

I

J

K

L

M

T

V

W

Z

Serbian Astronomical Society
Volgina 7
11000 Belgrade
Serbia

ISBN 978-1-5108-7972-0

9 781510 879720

CURRAN ASSOCIATES INC.
proceedings.com